Geology of the Henry Mountains, Utah,
as recorded in the notebooks of G. K. Gilbert,
1875–76

Grove Karl Gilbert, 1843–1918

The Geological Society of America
Memoir 167

Geology of the Henry Mountains, Utah, as recorded in the notebooks of G. K. Gilbert, 1875–76

Charles B. Hunt
2131 Condie Road
Salt Lake City, Utah 84119

1988

Published by The Geological Society of America, Inc.
3300 Penrose Place, P.O. Box 9140, Boulder, Colorado 80301

Printed in U.S.A.

GSA Books Science Editor Campbell Craddock

Library of Congress Cataloging-in-Publication Data

Gilbert, Grove Karl, 1843–1918.
    Geology of the Henry Mountains, Utah, as recorded
in the notebooks of G. K. Gilbert, 1875–76.
    (Memoir / Geological Society of America ; 167)
    Includes index.
    1. Geology—Utah—Henry Mountains.   2.Intrusions
(Geology)—Utah—Henry Mountains   I. Hunt, Charles
Butler, 1906–   .   II. Title.   III. Series: Memoir
(Geological Society of America) ; 167.
QE170.H4G55   1988   557.92'52   88-2614
ISBN 0-8137-1167-3

Contents

INTRODUCTION

by Charles B. Hunt

Overview of the notebooks: Why Gilbert went to the Henry Mountains. Getting there, more than 200 miles from Salt Lake City, 20 days travel, mostly by pack train; use of prehistoric trails for finding water and routes to passes and passable canyons. Problems with names.

While learning the stratigraphy along the route to the mountains, Gilbert discovers two stages of structural deformation, now known as Laramide and middle Tertiary.

Lavas or intrusions? Discovering that the "lavas" actually are intrusions leads to discovering the form of the intrusions and the fact that intrusions deform the host rocks into which they are intruded.

Erosion, transportation, and corrasion. A third of Gilbert's monograph is devoted to land sculpture.

Surveying the topography, the first one-degree topographic quadrangle maps of that area.

References cited and some related ones.

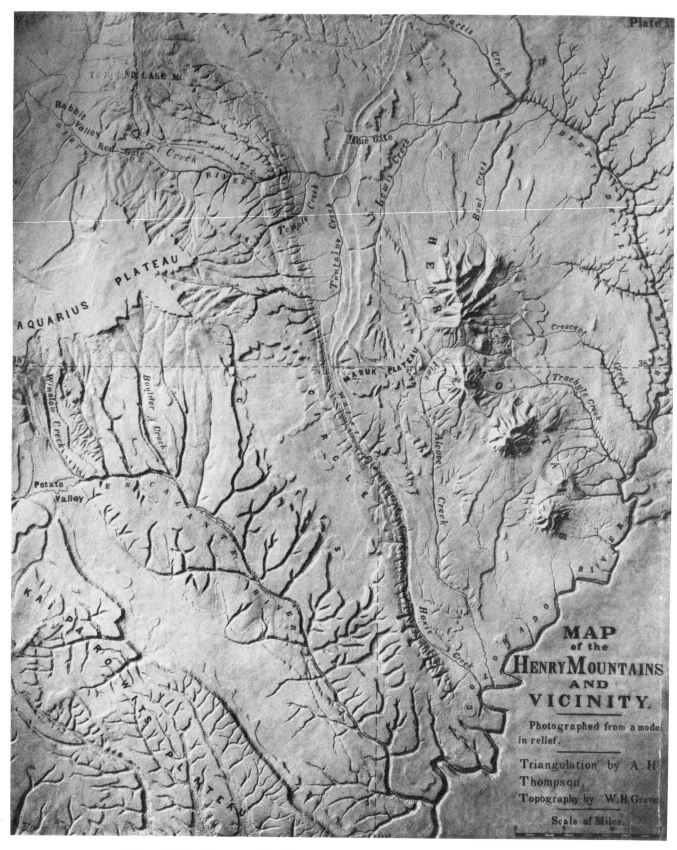

Figure 0.A. Landform map of the Henry Mountains and vicinity (from Gilbert, 1877b, pl. 1).

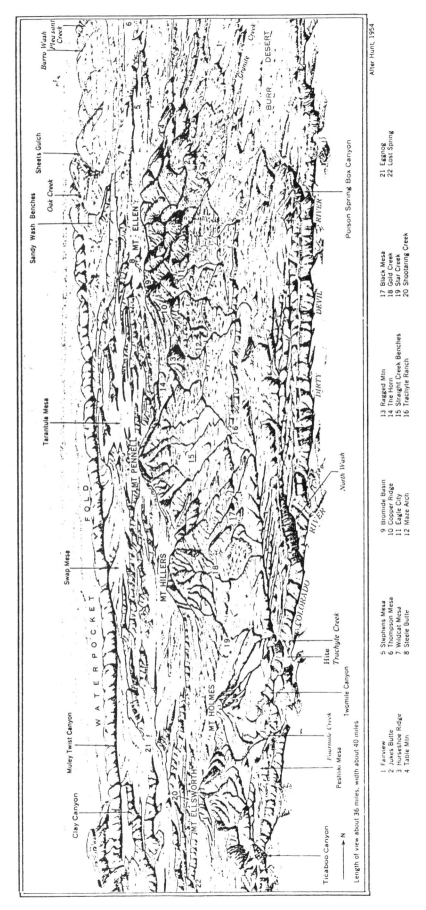

Figure 0.B. Diagrammatic view of the Henry Mountains region (from Hunt, 1956, Fig. 2. The original, in U.S. Geological Survey Prof. Paper 228, pl. 2, extends north to the San Rafael Swell and south to where the Colorado River crosses the Waterpocket Fold).

1 Fairview
2 Jukes Butte
3 Horseshoe Ridge
4 Table Mtn

5 Stephens Mesa
6 Thompson Mesa
7 Wildcat Mesa
8 Steele Butte

9 Bromide Basin
10 Copper Ridge
11 Eagle City
12 Maze Arch

13 Ragged Mtn
14 The Horn
15 Straight Creek Benches
16 Trachyte Ranch

17 Black Mesa
18 Gold Creek
19 Star Creek
20 Shootaring Creek

21 Eggnog
22 Lost Spring

OVERVIEW OF GILBERT'S NOTEBOOKS

Gilbert's field notes about Utah's Henry Mountains are far more than a day-by-day account of his activities and observation. Reading them in the context of the status of knowledge about geological science at that time, 1875–1876, one quickly finds that the notes are an exciting record of what were totally new scientific discoveries representing milestones—breakthroughs—in the history of geological knowledge. One experiences the thrill of reading the firsthand account of the discovery of several concepts that today are accepted routinely as fundamental in the science.

Grove Karl Gilbert, 1844–1918, was one of the greatest geologists America has produced. Many share my opinion that he was the tallest of the several giants who contributed so much to the healthy development of the young science. His contemporaries evidently thought so also, because twice they elected him President of the Geological Society of America, the only person to be so honored. His contributions, of course, are recorded in his published monographs, but his field notes provide an inside view of how his major discoveries developed.

WHY GILBERT STUDIED THE HENRY MOUNTAINS

John Wesley Powell, in his trips down the Colorado River (1869–1871), had a good view of the two southern Henry Mountains (Fig. O.C) and saw that they were structural domes associated with "lavas." Geologists of that time still were debating whether volcanoes were "craters of elevation" (von Buch, 1836) or merely piles of accumulated lava and other ejecta around the craters. Powell could see that the two mountains were domed, for the massive Triassic-Jurassic sandstones conspicuously rise on their flanks. The "lavas" could be seen as dark masses at the centers of the domes, and specimens could be examined in the drainage courses off the mountains. Surely these were craters of elevation.

And so Powell arranged for Gilbert to visit the Henry Mountains and determine the facts. To reach the Henry Mountains, Gilbert had to cross the lavas of the High Plateaus, from Salina Canyon south to Fishlake and around Rabbit Valley east of Fishlake. From a distance, and for quite a while after being in the Henry Mountains, Gilbert continued to accept the idea that likewise the igneous rocks in the Henry Mountains are lavas and referred to them as such in his first notebooks.

GETTING THERE

Most of southeastern Utah and all of the Henry Mountains area was a vast, unmapped, vacant wilderness during the 1870s when Gilbert made his trip there. Of the Spaniards who explored the southern part of the Colorado Plateau in the 16th century, only Cardenas saw the Colorado River—at the Grand Canyon in 1540. Only Escalante, in 1776, crossed the river. Ives (1861) was the first of the U.S. exploratory surveys to describe the region. He wrote discouragingly, "It seems destined by nature that the Colorado River, along the greater part of its lonely and majestic way, shall be forever unvisited and undisturbed." This still seems to be the view of Utahns, despite the fact that southeastern Utah's unparalleled scenery is potentially the state's greatest resource.

Other exploratory surveys east of the Colorado River while Gilbert was visiting the Henry Mountains could have seen the Henry Mountains only as unidentified peaks. Holmes' interpretation of the intrusive structure at Ute Mountain (El Late) anticipated Gilbert's interpretation of the Henry Mountains. In fact, Holmes even recognized that the bulging intrusions there are linear and radiate from a center. And Gilbert surely was mindful of the fact that Peale's attempt to study the La Sal Mountains had to be aborted because of hostility of the Indians.

The Henry Mountains did not appear on any map until Powell discovered them in 1869. Powell named them for Joseph Henry, physicist at the Smithsonian Institution who assisted Powell in obtaining financial support for his exploration of the Colorado River. Even as recently as the 1930s, when we made our survey of the region (Hunt, Averitt, and Miller, 1953; publication was delayed by World War II), it still was the center of an area the size of New York State without a railroad (and it still is), and a third of that was without a road of any kind. The Henry Mountains and the country around them still was pack-train country, and to work there provided an opportunity to share the experiences of pioneers like Gilbert and his party who made the first maps. Ours was the last of the big pack-train surveying projects that had begun in the 1850s as surveys for a transcontinental railroad—the end of an era.

How did Gilbert find his way across thousands of square miles of vacant, unmapped, rugged wilderness to reach the Henry Mountains, and how was he able to always complete a day of travel and end at a spring? He did so by following the prehistoric Indian trails. So also did the frontiersmen who explored for grazing or farm lands and the three parties of prospectors who had preceded him to the Henry Mountains (Gilbert, 1877b, p. 151). To explorers like Gilbert, those Indian trails were as easy to follow as a modern freeway. They were the conspicuous routes through the otherwise trackless wilderness before there was grazing by livestock. Moreover, one can even find directional signs along the trails indicating whether one is approaching water or going away from it. The secondary trails of coyotes, foxes, and rabbits diverge from the main trail away from water; they converge toward water. These signs of the past still are preserved in Death Valley where there has been no grazing (Hunt, 1975, p. 175–178). That Gilbert knew these signs and could use them to his advantage is indicated by the ease with which he generally found satisfactory camp sites that provided the pasturage and water for his animals as well as his men. In two field seasons, only a couple of his camps proved unsatisfactory.

Figure 0.C. Oblique air photograph of Mts. Holmes and Ellsworth showing canyon-forming sandstones (Wingate, Kayenta, and Navajo sandstones) domed by the "lavas" (dark rocks) at the central parts of the mountains. To Powell, these assuredly were craters of elevation, and so he arranged for them to be studied by Gilbert. In the distance is the Waterpocket Fold (Gilbert's Escalante fold) and the Circle Cliffs uplift. On the skyline left is the Kaiparowits Plateau. Right skyline is the Aquarius Plateau. (Photograph by Fairchild Aerial Surveys.)

To considerable degree this still was true of the Henry Mountains area during the 1930s, and the geologists had to learn to distinguish between the "big" trails and the "dim" ones (Hunt, Averitt, and Miller, 1953, p. 12). The ancient Indian trails had by then been destroyed by livestock, but the cows and wild horses marked the routes to and from water.

Gilbert knew he was pioneering in exploring the Henry Mountains area, and his published report begins: "The Henry Mountains have been visited only by the explorer" (Gilbert, 1877b, p. 1). He further wrote (p. 14), "No one but a geologist will ever profitably seek out the Henry Mountains . . .," but he reveals his enthusiasm for the area by following this statement with three or four pages of discussion of how to get there. "At Salt Lake City

he can procure pack-mules and pack saddles, or 'apparejos' and everything necessary for a mountain outfit."

Distances.

From Salt Lake City to Salina	155 miles
From Salina to Fish Lake	38 miles
From Fish Lake to Rabbit Valley	27 miles
From Rabbit Valley to Temple Creek Cañon [*Pleasant Creek*]	27 miles
From Temple Creek Cañon to Lewis Creek [*Sweetwater Creek*]	18 miles
Thence to base of Mt. Ellen	10 miles

(Gilbert, 1877b, p. 17)

And Gilbert further reveals his enthusiasm for the geology in that area by recording in his notebook 4 (p. 36) that both the San Rafael Swell and Circle Cliffs uplift "are worthy of monographs."

PROBLEM WITH NAMES

After Gilbert left Salina, he entered a country without maps and almost totally without local names, either for places or for formations. Even Mt. Ellen, largest of the Henry Mountains, had to be referred to as Henry I during the first part of his field work. The smallest of the five mountains did not become known as Mt. Holmes until after his field work was completed. In the field notes this is Henry V. Gilbert had to create a set of field names both for the places he wanted to describe and the formations that were part of them. To keep track of the places that were of interest geographically as well as geologically, Gilbert made landscape drawings and assigned numbers to places he wished to record. In his notes he frequently speaks of a location by its number. This makes reading difficult, and it seemed desirable to list the names and numbers that he used and to identify them by their present-day names (Table 0.1). Additionally, I have added the identifying names in brackets (indicating this editor's addition) for some of the important names, especially in the early chapters of the book. The table brings out the fact that many of the names used by Gilbert in his published report differ from his field usage, so there is a column for each of those usages.

To considerable degree there was confusion in place-names as recently as the 1930s. Persons who used the mountains during the summertime, for grazing or for prospecting, had one set of names. Those who used the desert and canyons during the cold months had a different set of names. The names on the maps accompanying U.S. Geological Survey Professional Paper 228 (Hunt, Averitt, and Miller, 1953) attempt to resolve the differences. But still a new and different set of names was introduced during the uranium boom of the 1950s. The 15-minute quadrangle maps published during that period attempt to resolve conflicts in nomenclature to that time.

Examples of present-day confusion in place-names are numerous. For instance, Trochus Butte of Gilbert and of USGS Professional Paper 228 has become Turkey Knob of the USGS Bull Mountain quadrangle map; Bulldog Peak on the east side of South Pass in USGS Professional Paper 228 has become Cass Creek Peak on the Mt. Hillers quadrangle map and the name Bulldog applied to the hilltop west of the pass; Sawtooth Ridge in Professional Paper 228 has become Cocks Comb on the Mt. Hillers quadrangle. Asking for directions in the Henry Mountains area can be difficult.

Changes in the geographic names in the Henry Mountains area even record changes in the mores of our culture. Some local people had objected to the name Dirty Devil for the principal stream in the area. When the Board of Geographic Names was told by a postmistress that "the nicer people call it Fremont," the name Powell had given to the stream was dropped. Because of its history, I was interested in restoring Powell's name. It even had inspired the name "Bright Angel" for the very different stream in Grand Canyon. When I found that most local people still used the name "Dirty Devil," I related the earlier protest and was told, "Oh, the nicer people moved away from here a long time ago."

So the Fremont and the Muddy rivers come together to form the Dirty Devil.

How little was known about that country in Gilbert's time is indicated by the fact that Powell's attempt to have supplies brought to the mouth of the Dirty Devil in 1871 failed because the supply party mistook the Escalante for the Dirty Devil. A year later, Professor Thompson (1939, and in Powell, 1872) discovered the mistake and retrieved the boat that Powell had cached near the mouth of the Dirty Devil.

LAVAS OR INTRUSIONS?

Gilbert's ideas about the structural forms of intrusions, notably the bulging forms in the Henry Mountains that he later called lacunes, then laculites, and finally laccolites, have been summarized elsewhere (Hunt, 1980). Reading Gilbert's notebooks, however, drives home the fact that before he could develop his ideas about the form and mechanism of intrusion, he had to discover that the igneous rocks in the mountains are not extrusives—lavas. This breakthrough is recorded by his change in nomenclature of the igneous rocks. During his 1875 visit the igneous rocks are referred to as lavas. During 1876 they are referred to as trap.

Page 51 of Notebook 3 has an unlabeled cross section of Gilbert's classical laccolith (Fig. 3.51). It is the first such cross section. But on page 15 of the next notebook he still writes, "I have now seen the lavas of all the 5 Henries." The idea that the igneous rocks are intrusive had started to grow on his first trip and seems to have matured during the winter between his two trips to the Henry Mountains.

Discovery that the "trap" is intrusive clarified the structural geology he found in the Henry Mountains. Now the puzzling structural bulges could be explained, and in one entry (Notebook 4, p. 2) he refers to "Bubble Mountains." [see also p. 3-54 and 3-59] He could see that the individual masses were separate intrusions and that each had deformed the sedimentary formations that were intruded. Today this seems simple, straightforward, and obvious, but the notebooks reveal that Gilbert felt some excitement about his discovery. His evidence that the rocks are intrusive and not extrusive lavas was conclusive and is outlined in full in his monograph (Gilbert, 1877b, p. 51–53; see also quotation in full in Hunt, 1980, p. 28). Gilbert's report was the first to clearly point out that intrusive igneous rocks deform their host rocks. The notebooks reveal how this major breakthrough was achieved.

In his monograph, Gilbert summarized the evidence for intrusion as follows (Gilbert, 1877b, p. 51–54):

1. No fragment of the trachyte has been discovered in the associated strata. . . .
2. The trachyte is in no case vesicular, and in no case fragmental. . . . [*This is true at the laccoliths, but the stocks are bordered by shatter zones having fragmental blocks of trachyte mixed with the fragmented sedimentary rocks.*]

TABLE O.1. MODERN NAMES OF PLACES AND FORMATIONS FREQUENTLY REFERRED TO BY GILBERT*

Gilbert's Field Notebooks	Gilbert's Published Report	Modern Names**
A Cliff, see under Cliff		
Aubrey Sandstone	Aubrey Sandstone	White Rim, Cedar Mesa, and other upper Permian sandstones
B Cliff, see under Cliff		
Belted Cliff	Henry's Fork Group	Morrison Formation and underlying Summerville Formation
Blue Gate, B.G. Sandstone, and Blue Gate Shale	Blue Gate, Blue Gate Sandstone,	Blue Gate is the valley of the Fremont River between the North and South Caineville Mesas; Emery Sandstone caps the mesas; Blue Gate Shale forms the slopes
Button	--	Isolated fault block of lava along the west edge of Boulder Mountain, about 100 feet lower than the top of the plateau and about one-half mile west
C Cliff, see under Cliff		
Cache Creek	--	Dugout Creek
Camp Cache	--	Mouth of Dugout Creek at west base of Mt. Ellen
Circle Cliffs	--	Circle Cliffs
Cliff A	*Gryphaea* Sandstone	Dakota Sandstone
Cliff B	Tununk Sandstone	Ferron Sandstone
Cliff C	Blue Gate Sandstone	Emery Sandstone
Cliff D	Masuk Sandstone	Mesaverde Formation
Corral Creek	Corral Creek	Sulphur Creek
Crescent Creek	Crescent Creek	Crescent Creek, joined by its South Fork out of Maze Arch forms North Wash. Called Pleasant Creek in notebook 4
Dinah Creek, Pass	Dinah Creek, Pass	South Pass, between Hillers and Pennell
D.D., see Dirty Devil		
Dirty Devil fold	--	Faulted south flank of the Teasdale anticline
Dirty Devil River	Dirty Devil River	Name now restricted to the segment below the junction of the Fremont and Muddy Rivers
Escalante fold	Waterpocket fold	Waterpocket Fold
Gate Sandstone	Blue Gate Sandstone	Emery Sandstone
Gate Shale	Blue Gate Shale	Blue Gate Shale
G.C. (Gray Cliff) Sandstone	Gray Cliff Sandstone	Navajo Sandstone
G.M., Great Massive	Gray Cliff Sandstone	Especially the Navajo Sandstone, but at places may include other formations of the Glen Canyon Group
Gryphaea bed	*Gryphaea* Sandstone	Dakota Sandstone
H.I. = Henry I	Mt. Ellen	Mt. Ellen
H.V.	Mt. Holmes	Mt. Holmes, smallest of the five Henry Mts.
Hilloid Butte	Marvine laccolite	Table Mountain bysmalith
Howell Fold	--	Capitol Reef

TABLE 0.1. MODERN NAMES OF PLACES AND FORMATIONS FREQUENTLY REFERRED TO BY GILBERT* (continued)

Gilbert's Field Notebooks	Gilbert's Published Report	Modern Names**
Hoxie Creek	Hoxie Creek	Hall Creek
Jerry	Jerry Butte	Sawtooth Ridge extending east from Mt. Hillers
J.T., Jura-Trias	--	Jura-Trias
l.b., Lower Blue	--	Tununk Shale
lacune	laccolite	laccolith
Leaden	--	Ferron or Emery Sandstone south of Tarantula Mesa
Lower Gate Cliff	Blue Gate Shale	Blue Gate Shale
Lower Gate Sandstone	Blue Gate Sandstone	Emery Sandstone? Possibly Ferron Sandstone
Maze Arch	Maze Arch	Maze Arch, anticlinal uplift along east base of Mt. Ellen
M.L.	Thousand Lake Mtn.	Thousand Lake Mtn. M = 1000
NE Butte	Jukes laccolite	Bull Mountain
Needle Butte	--	Factory Butte
North Twin Mesa	--	North Caineville Mesa
Oyster Bed	_Gryphaea_ Bed	Dakota Sandstone
Pass Butte	Sentinel Butte	The Horn
Pine Alcove Creek	Pine Alcove Creek	Bullfrog Creek
Red Arena	--	Dome exposing Jurassic formations north of west of the Blue Gate
Red Gate	Red Gate	Valley of Fremont River where it is incised into the Jura-Trias formations of the Teasdale anticline east of the Bicknell Bottoms
Round Top Mesa	--	Miners Mountain
S.R.	San Rafael	San Rafael
Sentinel Butte	Sentinel Butte	The Horn
Shin	Shinarump Group	Includes Chinle Formation, Shinarump Conglomerate, and Moenkopi Formation
South Cache Creek	--	South Creek
South Canyon of Hillers	--	Star Canyon
Square Top	--	Lion Mesa
Summit Spring	--	Spring in Penellen Pass
Supply Camp	--	South of Fremont River at the Red Gate
Tantalus Creek	Tantalus Creek	Sand Wash
Temple Creek	Temple Creek	Pleasant Creek
Twin Mesas	--	North and South Caineville Mesas
Upper Blue Cliff	Blue Gate Sandstone	Emery Sandstone
Upper Lead Cliff	Masuk Sandstone	Mesaverde Formation
Var., Variegated beds	Upper part of Shinarump Group	Chinle Formation
Vag., Vagabond beds	Flaming Gorge Group	Entrada Sandstone and upper part of Carmel Formation

TABLE O.1. MODERN NAMES OF PLACES AND FORMATIONS FREQUENTLY REFERRED TO BY GILBERT* (continued)

Gilbert's Field Notebooks	Gilbert's published report	Modern names**
Ver., Verm., Vermilion Sandstone	Vermilion Cliff Group	Wingate Sandstone and lower part of the Keyenta Formation
Very South Cache Creek	--	South Fork of South Creek
Waterpocket Canyon	Waterpocket Canyon	Canyon of Hall Creek where incised into the Waterpocket Fold

*Other names infrequently used may be found in the index. For camps, instrument stations, and points sighted, see appendix.
**From U.S. Geological Survey Professional Papers 164, 228, and 363 and/or the 1:62,500 topographic quadrangle maps.

3. The inclination of the arched strata proves they have been disturbed....
4. It occasionally happens that a sheet, which for a certain distance has continued between two strata, breaks through one of them and strikes across the bedding to some new horizon, resuming its course between other strata....
5. The strata which overlie as well as those which underlie laccolites and sheets, are metamorphosed in the vicinity of the trachyte, and the greatest alteration is found in the strata which are in direct contact with it....

He further notes that

1. ... seven laccolites ... lie so far above the local plane of erosion ... they have no enveloping strata, and their only associated sheets lie in the strata under them....
2. There are two laccolites so nearly bared that their forms are unmistakable, but which are still partially covered by arching strata, and which have associated sheets and dikes....
3. There are five supposed laccolites situated where the erosion planes are inclined, which run under the slopes at one side or end, and at the other project so far above them as to have lost something by erosion....
4. There are seven or eight supposed laccolites, of which only a small part is in each case visible, but which are outlined in form by domes of overarching strata. Their bases are not exposed....
5. There are five domes of strata accompanied by dikes and ... sheets but showing no laccolite....
6. There are nine or more domes of strata with no visible accompaniment of trachyte....

He then discusses how the laccolite hypothesis explains all these different kinds of occurrences.

Despite the overwhelming and convincing evidence that Gilbert marshalled about the intrusive origin of the igneous rocks in the Henry Mountains, many geologists were not convinced. Reyer (1888, p. 135), for example, insisted that the laccoliths must be surface lavas buried by the overlying sedimentary formations. Neumayr (1887, p. 180) stated that the evidence presented by Gilbert was very convincing but so surprising that further confirmation was needed.

Also, while Gilbert was in the Henry Mountains, contemporaries working with the Fortieth Parallel Survey were examining the stocks at Park City and Cottonwood canyons in the Wasatch Range and concluding that those stocks were protuber-

ances of Precambrian granite overlapped by the sedimentary formations around them (Hague and Emmons, 1877, p. 353). The issue still was being debated at the International Geological Congress meeting in Salt Lake City during the 1890s. Gilbert did not comment, in writing, about their interpretations. He simply waited for the issue to be resolved by the facts he had discovered at the Henry Mountains and facts being discovered elsewhere.

Gilbert's conclusions about the intrusive origin of the igneous rocks in the Henry Mountains were enhanced by the relations he found at the sills and dikes southwest of the San Rafael Swell. These observations, given in Notebook 10, provided additional criteria for identifying intrusions.

REGIONAL STRUCTURE AND STRATIGRAPHY

Before Gilbert could decipher the structural geology of the Henry Mountains, he had to develop knowledge about the stratigraphy of the late Paleozoic, Mesozoic, and early Tertiary formations in the area. In this he used his paleontological training. He found all the important fossil-bearing beds and collected from them. His notes record many of his own identifications as to genera and species. His attention to the paleontological detail enabled him to correctly identify the ages of the formations.

He did this while traveling there, and while learning the stratigraphy he discovered that the region had undergone a succession of displacements; an early set of displacements that we now call Laramide produced the Kaibab type of structure (Circle Cliffs, San Rafael Swell, and others). These displacements were followed by regional uplift and block faulting during the middle and late Tertiary (see for example his pages 26 and 51 in Notebook 4). Commuting by pack train was slow but had its advantages, and Gilbert's notes show that getting there was half the fun.

Even today, many geological reports have been written as if all the structural displacements on the Colorado Plateau were Laramide, a badly misused term. Gilbert not only showed otherwise but also recognized that the displacements took place slowly over a long period of time and that dissection by the streams progressed pari passu with the displacement (Notebook 1, p. 58).

Gilbert was puzzled by the Maze Arch, the anticline paralleling the elongation of Mt. Ellen and located along the east base

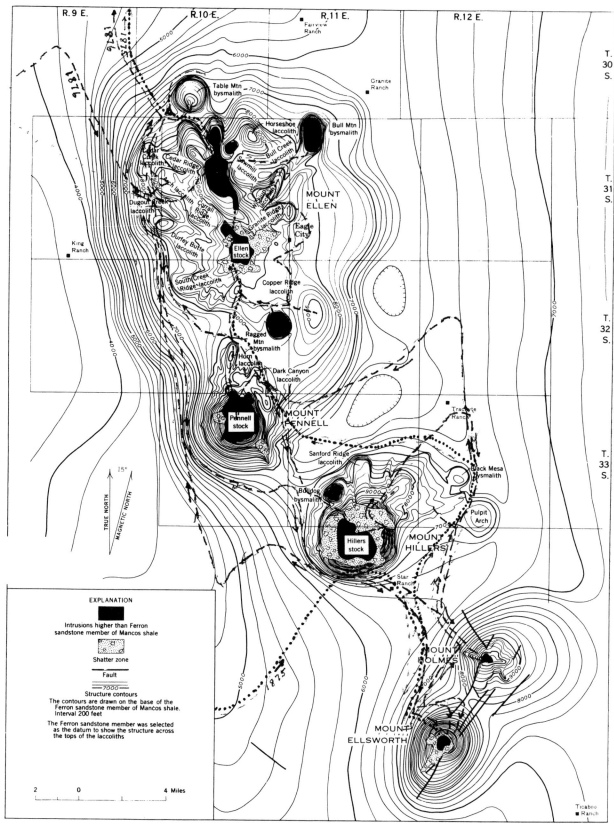

Figure 0.D. Structure contour map of the Henry Mountains with Gilbert's routes—1875 dotted, 1876 dashed—showing the intrusions on which he based his monograph. (Map from Hunt, Averitt, and Miller, 1953.)

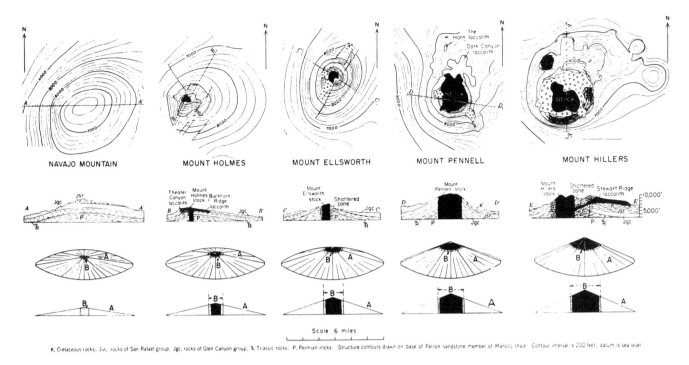

NAVAJO MOUNTAIN   MOUNT HOLMES   MOUNT ELLSWORTH   MOUNT PENNELL   MOUNT HILLERS

K, Cretaceous rocks, Jsr, rocks of San Rafael group, Jgc, rocks of Glen Canyon group, Tr, Triassic rocks, P, Permian rocks. Structure contours drawn on base of Ferron sandstone member of Mancos shale. Contour interval is 200 feet, datum is sea level

Figure 0.E. Structure contour maps and cross sections of 4 of the Henry Mountains and of Navajo Mountain showing stages in the development of the stocks and of the big mountain domes they produced. Additional mapping of the structural geology at other laccolithic mountains on the Colorado Plateau has greatly increased the number of stages that are represented by the series. Mt. Ellen is irregular and does not fit the series (see also Affleck and Hunt, 1980, p. 110; from Hunt, 1942).

of the mountain (Fig. O.D). We still are puzzled. Its structural form is similar to the uplifts at the Kaibab, Circle Cliffs, and San Rafael Swell, but by comparison it is minute. There is not another fold like it on the Colorado Plateau. And inspection of the structure contour map or of the magnetic maps (Affleck and Hunt, 1980) reveals no hint of relation to the igneous structures.

THE INTRUSIVE FORMS

Our survey during the 1930s (Hunt, Averitt, and Miller, 1953) abundantly confirmed Gilbert's conclusion that the igneous rocks in the Henry Mountains are intrusive. Indeed, our survey has been described as having determined in five years what Gilbert discovered in two months.

Because of the resurvey, the structure contour map that could be constructed (Fig. O.D) brought out that the laccoliths are linear, tongue-shaped bulges producing anticlines that radiate from the stocks, rather than circular domes as Gilbert had inferred. His routes, shown approximately on the structure map, emphasized the rounded distal ends of the intrusions. Yet Gilbert even anticipated our mapping of the shapes by noting (1877b, p. 38) that some of the laccoliths "jut forth . . . like so many dormer windows."

He did not identify the stocks as crosscutting, steep-walled intrusions, as is indicated by their mappable crosscutting walls (from Permian to Cretaceous at Mt. Hillers) and by the more recently mapped negative magnetic anomalies north of the positive ones over the stocks (Affleck and Hunt, 1980). The big stocks are not well exposed; they had to be discovered by mapping. Gilbert did not have time to study those areas so missed seeing their increased differentiation, increased metamorphism, indicated higher temperatures, signs of metallization (see Notebook 9, p. 34), and peripheral shatter zones, all of which are markedly different from those of the laccoliths and bysmaliths. Mapping the stocks disclosed also an orderly relation between their sizes and the sizes of the big domes around them (Fig. O.E).

Gilbert wondered, as have those who followed him, why the laccoliths spread laterally as they did. Why didn't their feeders, now recognized as the stocks, continue to the surface to erupt as volcanoes? In fact, of all the laccolithic mountains on the Colorado Plateau, only one, the North LaSal Mountain, shows good evidence of having erupted as a volcano (Waters and Hunt, 1958). Gilbert attempted to explain the lateral spreading by hydrostatics, but this would require magma too nearly fluid. Few accept that interpretation, but we still are not able to satisfactorily explain the problem he addressed.

After Gilbert had discovered that the intrusive rocks were not lavas, he first referred to the intrusions as "bulges" and "arches," but near the middle of his 1876 trip, inferring they had the form of a mushroom—an idea that had occurred to him at the end of his 1875 season but was not discussed (Fig. 3.51)—he coined the term *lacune* for them (see p. 48 in Gilbert's Notebook 8 and Figs. 8.26 and 8.27). This term evolved to laccolite in his published report (Gilbert, 1877b, p. 19). Dana (1880) suggested further changing the term to "laccolith" because the ending "ite" was so generally used for designating kinds of rocks.

It seems appropriate to record here the origin of the term cactolith that was formally introduced and defined in USGS Professional Paper 228 (p. 151). It was intended to call attention satirically to the absurd nomenclature geologists were developing by applying new names to the infinite variety of shapes intrusions can form. The name cactolith and its definition started July 1939 at what may be called elegantly a luncheon seminar on the outcrop of that feeder to the Trachyte Mesa laccolith. As shown on plate 12 of the professional paper, the intrusion is about ¾ mile southwest of Trachyte Mesa; it is elongate parallel to and aligned with the bulge at the mesa and aligned with the stock 4 miles farther southwest. Attending that seminar were Drs. J. W. Gregg, Frank Schairer, Earl Ingerson, and E. F. Osborn of the Carnegie Geophysical Laboratory and Dr. N. L. Bowen, then at the University of Chicago. I took the notes. Dr. Gregg proposed the name because the intrusion's shape so resembled the woody structure of the cane cactus. The definition that evolved from the discussion described it as a "quasihorizontal chonolith composed of anastomosing ductoliths whose distal ends curl like a harpolith, thin like a sphenolith, or bulge discordantly like an akmolith or an ethmolith."

I personally think that Gilbert would have approved even though the *New Yorker* did refer the definition to their "How's that again" department.

AGE OF IGNEOUS ACTIVITY

Gilbert inferred that the Henry Mountain intrusions were part of the igneous activity that was so extensive on the High Plateaus of Utah during the middle Tertiary. I was unable to improve on his estimate of the age and for 50 years have assumed the intrusions were middle Tertiary. For a while it seemed that radiometric dating methods, particularly K-Ar dates, would confirm Gilbert's estimated age. Recently, though, multiple dating methods have suggested a late Cretaceous age for the laccolithic intrusions in Colorado (Cunningham, Naeser, and Marvin, 1977).

It seems difficult to believe that such similar intrusions and intrusion sequences could be developed from such different structural environments as those of the late Cretaceous and middle Tertiary. By the middle Tertiary the Colorado Plateau in eastern Utah, site of the Henry Mountains, had been uplifted at least 2 and more likely 3 miles higher than it had been during the late Cretaceous. The problem has been summarized elsewhere (Hunt,

1983, p. 29–32). Until the conflict in the radiometric dates is resolved, one can stay with Gilbert's guesstimate or accept the accordant multiple age determination, which would suggest that the laccoliths spread while squishy muds still covered the bottom of the Cretaceous sea.

However, the determination of age is further complicated by that block of basalt, assuredly Tertiary, that was found in the shatter zone on the northeast side of the Mt. Ellen stock (Hunt, Averitt, and Miller, 1953, p. 157; see also Picard, 1980, p. 99). We were baffled by this basalt because it was not part of the stock and because at that time the extent of the middle or late Tertiary basaltic dikes and sills was not appreciated. Now it appears probable that the basaltic block represents another occurrence of those basalts caught up in the uplift around the Mt. Ellen stock. It could signal that the episode of laccolithic intrusions on the Colorado Plateau was later than the episode of intrusion of the basaltic dikes and sills. Gilbert's notebook 10 provides detailed descriptions of some of those dikes and sills southwest of the San Rafael Swell.

STRUCTURE CONTOURING

Who first attempted to show geologic structures by contouring the deformed beds? I do not know who was first, but surely Gilbert was one of the early geologists to try that methodology. His Figures 2.26, 2.28, and 2.29 along the Straight Cliffs, and Figure 3.58 where the Colorado River encounters the Waterpocket fold, appear to be form lines representing structure contours on gently folded beds. Is this another Gilbert first?

LAND SCULPTURE

Gilbert's notebooks give no hint that a third of his monograph (Gilbert, 1877b, p. 99–150) would be a classic account—a textbook—about land sculpture. His many references to geomorphic problems reveal they were on his mind, but his notes have no systematic analysis like his analysis of the structural geology of the intrusions.

In his report he discusses erosion processes—weathering, transportation, and corrasion. And he discusses the effects of declivity, rock texture, and climate on the rates of erosion. His discussion considers transportation and declivity, transportation and water quantity, corrasion and transportation, corrasion and declivity, and declivity and quantity of water.

Under sculpture, the production of the various topographic forms, Gilbert considers declivity (Notebook 9, p. 5), the law of structure (the greater resistance of hard rocks to the erosion processes), the law of divides that causes stream profiles to steepen toward the watershed, the influence of climate—greater rainfall tending to produce rounded rather than angular forms, the badlands (Notebook 9, p. 41, 55) that seem to contradict the law of divides, and the tendency for dynamic equilibrium between the passive and active elements in the processes.

Among the topics discussed by Gilbert under his heading

Figure 0.F. View of the Henry Mountains region, looking about south. Badlands in Mancos Shale are extensive along both sides of Sweetwater Creek, flowing right to left in foreground. Pediments around the badlands are overlapped by alluvium along Sweetwater Creek and its tributaries. At left and right (middle distance) are benches and cuestas of Cretaceous sandstones. Between these areas and the foot of the mountains are dissected gravel-covered benches, being higher level pediments (Gilbert's "hills of planation") buried by gravels from the mountains. Mt. Ellen is center, with Table Mountain bysmalith at the north base and Bull Mountain bysmalith at the left. The peak right of Mt. Ellen is Mt. Pennell; Mt. Hillers is directly back of Mt. Ellen. The east dipping Waterpocket Fold (Escalante fold in Gilbert's notebooks) is at the extreme right (dipping light colored formations). The distant mountain is Navajo Mountain. (Photograph by Fairchild Aerial Surveys.)

"Systems of Drainage" is the problem of planation, the origin of the gravel veneered pediments around the bases of the Henry Mountains. As has been indicated elsewhere (Hunt, Averitt, and Miller, 1953, p. 190), "The Henry Mountains could be as much the type locality for pediments as for laccoliths" because every stage in the processes of planation that produced the pediments (Fig. 0.F) can be seen, and so can every stage in the later aggradation that deposited the gravels covering and preserving them.

Gravel-free pediments that develop near the base of the mountains become extended and undercut gravel-laden streams discharging from the mountains. There are examples of incipient diversion so imminent one can imagine diversion being caused by the next storm (as in Fig. 8.D). One can find diversions so recent that deposition of the gravel fan on the bared pediment is just beginning. Other examples illustrate more advanced fan construction where a fan has been extended and the depositional pattern of distributary channels becomes modified to braided drainage, and the still more advanced stage where that drainage becomes parallel. And this sort of continuous disequilibrium continues until there is another diversion. Gilbert describes an example in Notebook 8, p. 20.

Gilbert sought dynamic equilibrium, but the history of planation around the base of the Henry Mountains illustrates progress of continuous disequilibrium. The mountains illustrate this further, for they have become taller and more rugged—that is, they have become more youthful—as they have become older. Canyon cutting adjacent to the Henry Mountains area (and at Grand Canyon) also does violence to the concept of the geomorphic cycle. In his monograph, Gilbert notes that as climatic conditions are varied, the law of structure becomes modified. Hence the round crests on Mt. Ellen are compared to the angular crest of Mt. Holmes. These kinds of problems are analyzed comprehensively in his monograph; some of the factual data are from his notes but not the systematic analysis.

The contrast in weathering and erosion of the stocks provides another example of the dynamic equilibrium Gilbert sought between structure and process. The northern stocks are higher and more watered than the southern ones, and each is deeply cut with a large valley whose sides are mantled by colluvium. Are those canyons due to more vigorous weathering and erosion, or might they be due to greater shattering and easier erosion at the large stocks?

SURVEYING THE TOPOGRAPHY

The topographic mapping of the area that began with Gilbert's party led to issuance of four one-degree quadrangle maps. The sheets, each covering a degree of latitude and longitude or about 3,500 square miles, were at a scale of 1:250,000 or roughly 1 inch equals 4 miles. The mapping was started by the Powell Survey and completed by the U.S. Geological Survey.

To tie his work to that of the topographers, Gilbert took a good many bearings to prominent topographic features and measured angles between points sighted to assist in the triangulation. The kind of equipment that was available to him and to the topographers is not known. He mentions using a plane table so I assume he had some kind of alidade and he must have had some kind of transit for taking the supplementary bearings that are recorded. Bearings recorded in his notes commonly have subscripts or superscripts that I do not understand, nor did inquiry at the Topographic Division of the U.S. Geological Survey provide explanation.

The published quadrangle maps of the Powell Survey do not indicate the declination at the time, and it is not mentioned in Gilbert's notebooks. He may have used only magnetic north, because his directions commonly are skewed that way. For example, his view from Ellen Peak (his Notebook 6) gives the direction, but his north is east of true north, east is south of true east, south is west of true south, and west is north of true west.

Gilbert faithfully recorded the altitude reading on his aneroid barometer, but how he made use of the information, other than for measuring short intervals, is not known. His notes do not record his calibrating his readings with the vertical angle determinations by the topographers. He evidently found the readings useful, but how is a mystery.

FORAGE, WILDLIFE

Although Gilbert does not discuss the matter, it seems clear that there was ample grass and other forage for his pack and riding animals. He rode the same horse day after day, and he commonly made single-night camp stands, so his animals were capable of working every day. By contrast, during the 1930s, forage was exceedingly scarce because of drought, and the condition was exacerbated by severe overgrazing. During our survey, each geologist had to have two riding animals, and it was necessary to use them on alternate days. Even so, oats had to be packed in for supplemental feed. Gilbert, on the other hand, was able to ride his horse regularly and did not have to haul in supplemental feed. Our bad range condition was general through the west and not confined to the Henry Mountains area; it led to enactment of the Taylor Grazing Act, the first attempt to control livestock on public lands other than national forests.

Gilbert's notes make little mention of wildlife. There is no mention of coyotes, foxes, or rabbits, yet surely they were there. During the 1930s we frequently were serenaded by coyotes; now, overkill has brought silent nights to the mountains and surrounding desert. Surely Gilbert had the experience of being startled by a jack rabbit darting from under a bush and spooking him and his horse while they sauntered along a lonely trail thinking about something remote from the moment. The buzz of a rattlesnake can be equally startling. No such incident is mentioned.

Gilbert found fishes "small but adult" at pools along Hall Creek where it is incised in a cañon cut into a part of the Waterpocket fold (Notebook 3, page 40). Gilbert does mention a few beaver. He found some signs of beaver along the Fremont River (his Dirty Devil) where it emerges from the Capitol Reef, and wrote (Notebook 4, page 42), "Camp fire is built of wood cut down by beavers—for what purpose? They cannot build dams here."

The only mention of deer was on his way out of the Henry Mountains in 1876. "Elisha kills a deer at dinner camp. (yes two)." (Notebook 10, p. 35). Elk tracks were observed on Thousand Lake Mountain (p. 6-39). Tracks of mountain sheep were recorded at a few mountain sites. On the summit of Mt. Ellen, Gilbert was so impressed by the steepness of their trails, he measured them—28° (p. 4-22).

At the back of Notebook 2 is an unlabeled drawing of a scorpion. Neither scorpions nor rattlesnakes are abundant in the area, but they are present and they were in Gilbert's time.

WEATHER

Gilbert's notes contain frequent reference to wind directions, including their changes as days warmed, and to the effects of particular topographic situations. His interest in wind directions continued during his studies in the Lake Bonneville basin. One of his diagrams is reminiscent of the *New Yorker*. It shows four smelters along the front of the Wasatch Range near the mouth of a canyon. Smoke from three of the smelters is blowing left; smoke from one is blowing right.

Gilbert also records snow, freezing of camp water jugs and canteens, and, occasionally, rain. These, however, seem in the nature of experiences rather than a summary of conditions that would affect land sculpture.

In his report and in his notes he does mention the scarcity of commercial timber.

STELLAR GEOGRAPHY

Gilbert was knowledgeable about and interested in stellar geography. Figure 6.7 is a sketch he made of the Pleiades. His Lake Bonneville notes contain several references to the planets visible at the time. As noted in Hunt (1982, p. 15), "Camp life—much of it without bothering with tents because of frequent moves, combined with clear desert skies—provided a nightly display of the heavens. Such living can hardly fail to stimulate interest in stellar geography. Very probably these outdoor experiences while with the Wheeler and Powell Surveys and USGS contributed to the interest that later became manifest in his descriptions and interpretations about the moon."

The Henry Mountains area is scenic by night as well as by day. High altitude vapor and dust are minimal because of the rain shadow cast by the High Plateaus.

TWO EDITIONS OF GILBERT'S REPORT

Gilbert completed his manuscript in 1877, during the winter following his second trip to the Henry Mountains. The first edition of his report carries the dateline 1877, but because of delays in preparing the illustrations it was not published until 1879. A second edition, published in 1880, is little different from the first except for an appendix summarizing related studies on the Colorado Plateau—at the La Sals, El Late, La Platas, Sierra Carriso, and Abajo Mountains. He also summarizes findings of parties working at the laccolithic mountains on the northern Great Plains. All these, like the Henry Mountains, reveal additional evidence of the intrusive nature of their igneous rocks and deformation of the strata by the intrusions.

So far as describing the Henry Mountains is concerned, the second edition of Gilbert's report on the Henry Mountains is mostly verbatim from the first edition. Pages 1 to 13 are identical and could have been from the same typeset. Page 14 of the second edition omits my favorite illustration, the head of a mule captioned in the first edition as "Ways and means," a drawing that was not only a work of art but also expressed the reality of the situation. Was this deletion an economy move? Surely it cost more to reset the type than was saved by removing the figure. The drawing is from Gilbert's Notebook 2, Figure 2.39, where the mule is identified as "Lazarus, Duke of York." From that deletion to page 60, the first edition is identical to the second. Dutton's

report about the rocks, which follows in the first edition, was reorganized, and the description of the rocks transferred to a separate chapter near the end of the second edition. The descriptions are slightly changed, but no important changes appear in the second edition. No changes appear in the chapters describing land sculpture and economic geology.

NOTES ABOUT EDITORIAL MARKS

As with the book about Gilbert's Lake Bonneville notes, editorial changes have been kept to a minimum, just enough to be fair to Gilbert. After all, these are field notes that he thought he was addressing to himself. Editorial corrections have been restricted to very few obvious misspellings, punctuation, capitals, or grammar, the attempt being to err by making too few rather than too many such changes.

Parentheses and double quotation marks (" ") used in the text are Gilbert's. Editorial comments are set in [*square brackets and italicized*]. Occasional words or phrases that could not be deciphered, either in part or at all, are enclosed in single quotation marks, such as 'clinf' or '   '. Gilbert's notebook page numbers are shown at the left of the column in curly brackets: {p. 4}.

ACKNOWLEDGMENTS

This book really began during the 1930s when the U.S. Geological Survey supported my resurvey of the Henry Mountains. It was the Survey's last big pack-train project and provided me with the opportunity to participate in experiencing a past era. During that time I had access to Gilbert's notebooks, which then were in the Survey's files. Now they are preserved in the National Archives and are only available by going to Washington, D.C. Having Gilbert's notes was an inspiration and education for this young geologist, and I learned at least some of what could be achieved following in Gilbert's footsteps, drawing landscapes to supplement photography, for example. I have elsewhere reported what I learned from Gilbert about writing reports (Hunt, 1959). From Gilbert and the opportunity provided by the Survey, I found that writing can be done easily and be fun. I am grateful.

Ellis Yochelson encouraged my writing chapters about Gilbert's work in the Henry Mountains and in the Lake Bonneville basin for the Geological Society of America Special Paper 183, "The scientific ideas of G. K. Gilbert" (1980). He obtained, through the U.S. Geological Survey, microfiche of Gilbert's notebooks that enabled me to prepare the book about his work in the Lake Bonneville basin (1982) and this account of his work in the Henry Mountains. The microfiche have been stored with the Photo Library of the U.S. Geological Survey in Denver.

Geoffry Brandt, my grandson, helped by copying from the microfiche some of Gilbert's more complicated drawings. Andrew Godfrey, who has worked in the Henry Mountains and currently is with the Fishlake Forest Service Office at Richfield,

Utah, had many suggestions to offer helping to reconstruct Gilbert's route from Salina to the Henry Mountains.

Finally, I acknowledge the help of my wife, in the field while mapping the mountains during the 1930s, again with students during the 1960s, and still again during preparation of this manuscript.

BIBLIOGRAPHY

Affleck, J., and Hunt, C. B., 1980, Magnetic anomalies and structural geology of stocks and laccoliths in the Henry Mountains, Utah, in Picard, M. D., ed., Henry Mountains symposium: Salt Lake City, Utah, Utah Geological Association, p. 107–112.

Armstrong, R. L., 1969, K-Ar dating of laccolithic centers of the Colorado Plateau and vicinity: Geological Society of America Bulletin, v. 80, p. 2081–2094.

Chorley, R. J., and Beckinsale, R. P., 1980, G. K. Gilbert's geomorphology, in Yochelson, E. L., ed., The scientific ideas of G. K. Gilbert: Geological Society of America Special Paper 183, p. 132–135.

Cross, C. W., 1894, The laccolithic mountain groups of Colorado, Utah and Arizona: U.S. Geological Survey Annual Report 14, p. 157–241.

Cunningham, C. C., Naeser, C. W., and Marvin, R. F., 1977, New ages for igneous rocks in the Colorado Mineral Belt: U.S. Geological Survey Open-File Report 77-573, 7 p.

Dana, J. D., 1880, Gilbert's report on the geology of the Henry Mountains: American Journal of Science, 3rd ser., v. 19, p. 17–25.

Davis, W. M., 1924, Gilbert's theory of laccoliths [abs.]: Washington Academy of Sciences Journal, v. 14, no. 15, p. 375.

——, 1925, Laccoliths and sills [abs.]: Washington Academy of Sciences Journal, v. 15, no. 18, p. 414–415.

——, 1926, Biographical memoir, Grove Karl Gilbert: National Academy of Sciences Biographical Memoirs, v. 21, 5th memoir, 303 p.

Dutton, C. E., 1880, Report on the geology of the High Plateaus of Utah: U.S. Geographical and Geological Survey of the Rocky Mountains Region (Powell), xxxii, 307 p., atlas.

Eckel, E. B., 1949, Geology and ore deposits of the La Plata district, Colorado: U.S. Geological Survey Professional Paper 219, 179 p.

Emmons, S. F., 1903, The Little Cottonwood granite body of the Wasatch Mountains: American Journal of Science, 4th ser., v. 16, p. 139–147.

Engel, C. G., 1969, Igneous rocks and constituent hornblendes of the Henry Mountains, Utah: Geological Society of America Bulletin, v. 70, p. 961–980.

Gilbert, G. K., 1876, The Colorado Plateau as a field for geological study: American Journal of Science, July and August 1876, p. 1–27.

——, 1877a, Geological investigations in the Henry Mountains, Utah [abs.]: American Naturalist, v. 11, p. 447.

——, 1877b, Report on the geology of the Henry Mountains: U.S. Geographical and Geological Survey of the Rocky Mountain Region (Powell), 160 p.

——, 1880, Report of the geology of the Henry Mountains: U.S. Geographical and Geological Survey of the Rocky Mountain Region (Powell), 2nd edition.

——, 1896, Laccolites in southeastern Colorado: Journal of Geology, v. 4, p. 816–825.

Gilbert, G. K., and Cross, C. W., 1896, A new laccolite locality in Colorado with its rocks [abs.]: American Geologist, v. 17, p. 407–408 (1896). See also Science, n.s. 3, p. 714 (1896).

Gilluly, J., 1927, Analcite diabase and related alkaline syenite from Utah: American Journal of Science, 5th ser., v. 14, p. 199–211.

——, 1929, Geology and oil and gas prospects of part of the San Rafael Swell, Utah: U.S. Geological Survey Bulletin 806-C, p. 69–130.

Godfrey, A. E., 1980, Porphyry weathering in a desert climate, in Picard, M. D., ed., Henry Mountains symposium: Utah Geological Association Guidebook to the Henry Mountains, 1980, p. 189–196.

Graf, W. L., 1980, Fluvial processes in the lower Fremont River Basin, in Picard, M. D., ed., Henry Mountains symposium: Utah Geological Association Guidebook to the Henry Mountains, 1980, p. 177–184.

——, 1983, Downstream changes in stream power in the Henry Mountains, Utah: Annals of the Association of American Geographers, v. 73, no. 3, p. 373–387.

Gregory, H. E., and Moore, R. C., 1931, The Kaiparowits region, a geographic and geologic reconnaissance of parts of Utah and Arizona: U.S. Geological Survey Professional Paper 164, p. 161.

Hague, A., and Emmons, S. F., 1877, Descriptive geology: U.S. Geological Exploration of the Fortieth Parallel (King Survey), v. 2, 890 p.

Hohl, A., and Everitt, B. L., 1980, Surficial geology of Bull Creek basin, in Picard, M. D., ed., Henry Mountains symposium: Utah Geological Association Guidebook to the Henry Mountains, 1980, p. 186–188.

Holmes, W. H., 1877, Report on the San Juan district, Colorado: U.S. Geographical and Geological Survey Territorial 9th Annual Report, p. 237–276.

——, 1878, Report on the geology of the Sierra Abajo and West San Miguel Mountains: U.S. Geographical and Geological Survey Territorial 10th Annual Report, p. 189–196.

Hunt, C. B., 1942, New interpretation of some laccolithic mountains and its possible bearing on structural traps for oil and gas: American Association of Petroleum Geologists Bulletin 26, p. 197–203.

——, 1946, Guidebook to the geology and geography of the Henry Mountains region: Utah Geological Society Guidebook to the Geology of Utah, 51 p.

——, 1956, Cenozoic geology of the Colorado Plateau: U.S. Geological Survey Professional Paper 279, 95 p.

——, 1959, About writing reports; some tips from G. K. Gilbert: Journal of Geological Education, v. 7, no. 1, p. 1–3.

——, 1973, 30 year photographic record of erosion on a pediment in the Henry Mountains area, Utah: Geological Society of America Bulletin, v. 84, p. 689–696.

——, 1975, Death Valley, geology, ecology, archeology: Berkeley, University of California Press, 234 p.

——, 1977, Around the Henry Mountains with Charlie Hanks; Utah geology: Utah Geological Survey, v. 4, no. 2, p. 95–104.

——, 1980, G. K. Gilbert on laccoliths and intrusive structure, in Yochelson, E. L., ed., The scientific ideas of G. K. Gilbert: Geological Society of America Special Paper 183, 148 p.

——, ed., 1982, Pleistocene Lake Bonneville, ancestral Great Salt Lake, as described in the notebooks of G. K. Gilbert, 1875–1880: Brigham Young University Geology Studies, v. 29, part 1, 225 p.

——, 1983,, Development of the La Sal and other laccolithic mountains on the Colorado Plateau: Grand Junction Geological Society Guidebook, 1983 field trip, p. 29–32.

Hunt, C. B., Averitt, P., and Miller, R. L., 1953, Geology and geography of the Henry Mountains region: U.S. Geological Survey Professional Paper 228, 234 p, Igneous geology reprinted in Picard, 1980.

Ives, J. C., 1861, Report upon the Colorado River of the West: 36th Congress, 1st session, House Ex. Doc. 90.

Johnson, A. M., and Pollard, D. D., 1973, Mechanics of growth of some laccolithic intrusions in the Henry Mountains, Utah, I: Tectonophysics, v. 18, p. 261–309.

Neumayr, M., 1887, Erdgeschicte, Bf.1, Allegemeine geologie: Leipzig, Bibliographischen Institute, 634 p.

Peale, A. C., 1877, On a peculiar type of eruptive mountain in Colorado: U.S. Geographical and Geological Survey Territorial Bulletin 3, p. 550–554.

Picard, M. D., ed., 1980, Henry Mountains Symposium, Salt Lake City, Utah: Utah Geological Association Publication 6 (contains maps and most text of Hunt and others, 1953, which is out of print).

Pollard, D. D., and Johnson, A. A., 1973, Mechanics of growth of some laccolith intrusions in the Henry Mountains, Utah, II: Tectonophysics, v. 18, p. 311–354.

Powell, J. W., 1872, Survey of the Colorado River of the west: U.S. Congress Documents, 42nd Congress, 2nd session, House Miscellaneous Documents

173, 291 p.

Pyne, S. J., 1980, Grove Karl Gilbert: Austin, University of Texas Press, 306 p.

Reyer, E., 1888, Theoretische geologie: Stuttgart, Schweizerbort (E. Koch), 867 p.

Thompson, A. H., 1939, Diary of: Utah Historical Quarterly, v. 7, nos. 1, 2, and 3, 140 p. See also in Powell, J. W., 1872.

von Buch, M., 1836, Description physique des Iles Canaries: Paris, 342 p.

Waters, A. C., and Hunt, C. B., 1958, Origin and evolution of the magmas, *in* Hunt, C. B., 1958, Structural and igneous geology of the La Sal Mountains: U.S. Geological Survey Professional Paper 294-I, p. 348–355.

Witkind, I. J., 1964, Geology of the Abajo Mountains area, San Juan County, Utah: U.S. Geological Survey Professional Paper 453, 110 p.

Yochelson, E. L., 1980, ed., The scientific ideas of G. K. Gilbert: Geological Society of America Special Paper 183, 148 p.

CHAPTER 1

Chapter 1. Notebook 1. May 29–June 22, 1875, Washington, Chicago, Salt Lake City, and to York, a ranch by the railroad in northern Juab County. Accompanied by Clarence Dutton and by Bell, one of the topographers. To Nephi and Taylor's Ranch. June 23 arrive at Gunnison. June 26 to Salina. June 27–30 camp at Salina, becoming acquainted with stratigraphy and structure of Salina Canyon. July 1 to camp 2 at coal in Salina Canyon. July 2–5 studying the stratigraphy and faults while ascending Salina Canyon to its head. July 6 climbing Mt. Hilgard. July 7 march southwest toward Fish Lake, moraines along the way. July 8 to Fish Lake. July 9, to Rabbit Valley (camp 7). July 10–13 working from Supply Camp, a major base camp where supplies were stored, which seems to have been located just south of the Bicknell Bottoms where the Fremont River is ponded upstream from the Red Gate at the Thousand Lake Fault. July 14, exploring Thousand Lake Mountain (M.L.). July 15, southward into the Fremont River valley in the Teasdale Anticline. July 16, 17, travel along the faulted southwest flank of the Teasdale anticline to camp 12 (July 17) near present-day Grover.

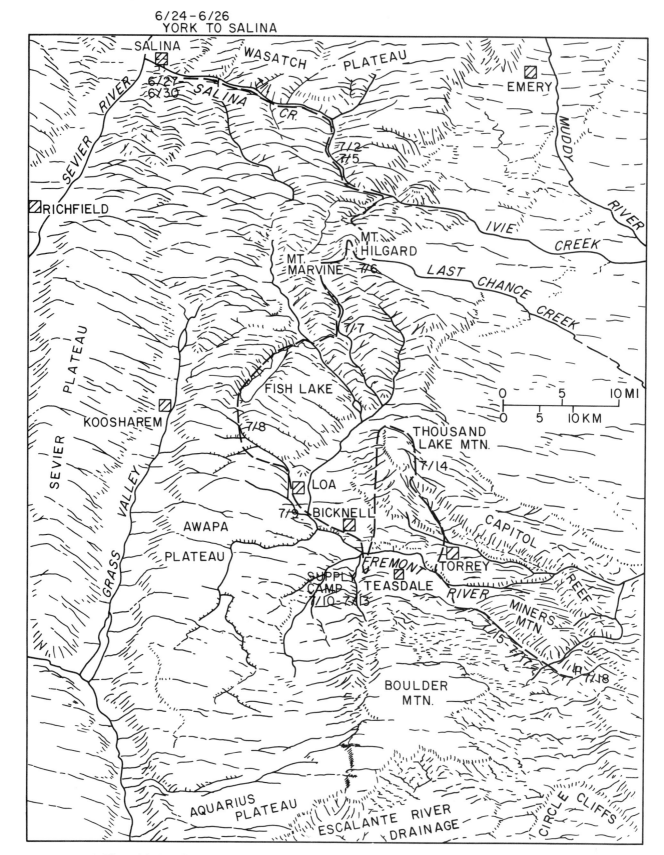

Figure 1.A. Landform map showing Gilbert's route Salina to Rabbit Valley, June 27 to July 17, 1875.

Figure 1.B. View southeast to Henry Mountains (about 50 miles) from pass between head of Ivie Creek and Thousand Lake Mountain. This is the first view Gilbert had of the Henry Mountains, but he does not refer to the sketch that he made (Fig. 1.19).

G.K. GILBERT, POWELL SURVEY, BOX 806, WASHINGTON, D.C.
NOTEBOOK NO. 1. OPENED AT YORK, UTAH. JUNE 20, 1875
CLOSED AT CAMP 12 NEAR SQUARE TOP BUTTE, UTAH, JULY 17, 1875.*

{p. 1}   **May 25, 1875.** Left Washington
June 13. Left Chicago
June 16. Arrived at Salt Lake City
June 20. Arrived at York by rail. Climbed a foothill [*of Mt. Nebo*] with Dutton and Bell, 1300 ft above camp. Found little

*For place names and Gilbert's abbreviations, see Table 0.1 and appendix.

rock in place. Pale gray limestone with concretions and a gastropod in one of them, supposed to be Tertiary. Lies near and probably immediately over a blue gray limestone with marine and probably Carboniferous fossils. *"Hemipronites"* and *"Lithostrotion?"* were found in local drift. A dolomite constitutes several ridges and is probably in dikes or massive eruptions, but at the surface it shows only boulders and no rock **in situ**.

Took charge of aneroid No. 3.

June 21. from camp 1 at York to camp 2 at Nephi.

The westward outlier of Nebo has an east dip and Nebo adjacent seems the same. But Nebo from the SW seems to dip W.

{p. 2}   **June 22.** Camp 2 at Nephi to Camp 3 at Taylor's ranch. Climbed Jurassic hills with Dutton. *"Pentacrinus and Nertina."* Appearances indicate the nonconformity between the Jura

Figure 1.C. View is south along the High Plateaus to Thousand Lake Mountain from near the same location as Figure 1.B.

and the red conglomerate (?) of the Mesa SE. The conglomerate dips east and is assumed to belong to the Gunnison synclinal. The Jura dips generally 30° to the E or SE. [*By Gunnison synclinal I assume Gilbert refers to the faulted syncline in San Pete Valley.*]

N of Salt Creek Cañon the same Jura shows but its relation to the adjacent spur of Nebo is not plain. The map of the spur shows a NW dip. Found boulders of trachyte but could not tell its source.

The Jurassic consists of shale, sandstone, and gypsum. The shale pale green and red, and sectile; weathers into a clay. The sandstone is in small force, is yellowish and contains many pebbles of shale. Only a single bed of gypsum seen but that 'firm' and thick 15–50 ft. An hour search discovered no fossils in the shale except trails, but the sandstone has numerous fragments of shells and shows obscure casts of conchifers that are not preserved.

On the march to Taylor's we had the Jura shown left all the way,

{p. 3}   and beyond it the Tertiary cliffs could be seen as far as Levan, where the Jura rises as high as to obstruct the view.

West of the valley the low range has an E dip—at least opposite Taylor's Ranch. I have not seen the beach line [*Bonneville Beach*] today, nor did I see it much south of York yesterday. We have ascended enough to make the water shoal and perhaps enough to cut it off.

[*The Lake Bonneville shoreline is indeed weakly developed in Juab Valley, the valley extending south from York. Short stretches of the beach are preserved between York and Nephi. The railroad station at Nephi is on the beach. It would cross to the west side of the valley near I-15, but in the middle of the valley it is*

Figure 1.1. Cross section on p. 4, Gunnison Ridge. a is the East Gunnison Ridge of '1 and 2 H'; b is red bad-land with gypsum and salt (Salt Cr.?) and runs conformably under d which is green gray and doubtless Jurassic; c is a trachyte flow; e is a lime and shale series (1.2.H?); f is the sub-non-com. series. Where I first touch it it's '   ' shale with some sandstone and gypsum in flakes.

Figure 1.3. Unlabeled section accompanying the description of Figure 1.2, p. 4.

buried under younger silt. On the west side of Juab Valley it is marked by lag concentrates of boulders and cobbles. Wave action was weak because, as Gilbert noted, the water shoaled.]
June 23. Camp 3 Taylor's to camp 4 Gunnison. 24.912 - AT 63. Bar 2144

Rode with Dutton among the foothills of the Sanpete Plateau and rejoined the road at the Sevier. The Jurassic almost disappears two or three miles south of Taylor's and its place is taken by the Tertiary. My observation is included by Howell's '   '.

June 24. In camp in Gunnison.

{p. 4}   **June 25.** Camp 4 in Gunnison.
June 26. From Gunnison to Salina with Dutton and Renshaw. Put up at Mr. Martin's. The same man who fed me when I was last here but not the same woman.

[A cross section sketched here, Fig. 1.1, is not identified. It seems to be Gunnison Plateau as seen from near Gunnison. On p. 5 is a cross section, Fig. 1.2, as sketched from Salina looking northeast. Fig. 1.3 is an unlabeled cross section, in with the explanation for Fig. 1.2.]

{p. 5}   *Sunday, June 27.* Up Salina Canyon at fault. Creek 25.35.

There are salt springs in the Jura close to this, and the honeycombed trachyte boulders in the arroyo attest to the solvent power of the salt. Some small pockets are dug to accumulate the salt by evaporating the water as it exudes.

Hilltop 24.17.
Looking northward (Fig. 1.4) the beds seen are t = trachyte covering the chief fault of the Jurassic area and 'thinning' into ridges by slides and perhaps faults. a = a pale gray under the trachyte and resting directly on the Jura. I supposed it to be volcanic ash. I do not see it east of the fault. j = Jura, within view almost completely red. Close to the fault it is interspersed with sandstone and this trends N and dips 70°E. (The fault trends 30° a a.) There seems no reason to doubt

{p. 6}   that the inferior beds E and W of the fault are adjacent portions of the same series, and as they are vertical on both sides, there is not much omission. T = Tertiary with a local dip of 12° which quickly diminishes to 5° a a.

Most of these features can be seen southward also. In that direction the strike of the Tertiary is 130° a a.
Depression to point John Thomas 5ᵃ ᵃ
"   " valley bluff   3ᵃ ᵃ

[Gilbert's recorded bearings include subscripts or superscripts like these. I am unable to explain them. See section on surveying topography in the Introduction.] On this hill the fault is locally marked by a calcite vein. In the sand bed west of this vein is a hole labelled "Kising 'Mn' lode."

Next I ascend the cañon up which I looked (N) from the hill. There are more extensive salt pools and a ruined salt boiler. In the pools the salt forms at the surface of the water and falls to the bottom in flakes. In the arroyos the salt crystallizes in coralloid forms like aragonite. There is much gypsum and silicate in flakes and crystals with salt '   '.

Figure 1.2. Sketch from Salina, looking NE. "a and b in Tertiary. c and d are Jurassic. The dip at a is about 15° (not measured). I think the two masses are separated by an E-W fault six miles N of here [*Fig. 1.3?*] whereby two phases of a monoclinal fold are revealed."

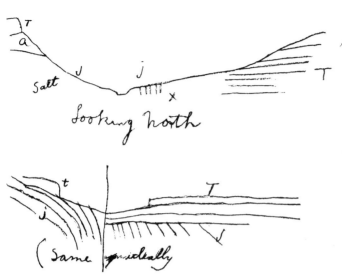

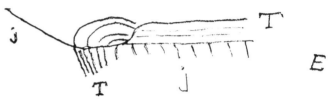

Figure 1.5. Section at x in Figure 1.4. a = red saliferous shale, no bedding visible; b = bastard limestone with chert; c = white calc shale, with a bed of bituminous shale with fish scales; d = chert; e = white calc shale; f = red shale; g = red sandstone; h = white sandstone and 'aren'-shale. Whole section about 250 ft.

Figure 1.4. Views of the N side of Salina Canyon, probably at the cross fault about 3 miles east of the mouth of the canyon. For section at x, see Figure 1.5.

Figure 1.6. Probable structure of the beds containing fish scales (=C? in Fig. 1.5).

{p. 7}   The section at *x* in Figure 1.4 is shown in Figure 1.5. The discovery of fish scales (see Fig. 1.5) leads to the surmise that this section is Tertiary and that the structure is that shown in Figure 1.6, the Tertiary being bent over till it dips 110°. The chert favors this and so do some of the phenomena of position.

However this is, the indisputable uncon. beds begin to show a half mile to the east. The general direction of the first part of the section is 70ᵃ ᵃ. The fish bed (1) [*in Fig. 1.7? previously described as bituminous shale as C in Fig. 1.5*] is identical in character with that which rests on its cut edges. [*It is not clear just where this section is located with respect to the preceding ones.*]

{p. 9.}   **Monday, June 28**

Stopped again in Salina last night and returned to the non-con rocks morning. The work last night stopped with bed 37. On the other side of the creek and within the horizon of No. 38 a coal seam has been found and opened. It is enclosed by tough clays and has a thickness of 15 inches. Strike SSW dip 30°E.

20 rods above the coal a fault shows on the south bank with a throw 20–30 ft to the east.

Section on p. 7–8 resumed (Fig. 1.8).

It is ½ or ¾ mile across the bar (unit 46 in Fig. 1.8) and riding to the

{p. 10}   E side of it I can see beyond no more of the lower levels. The red cap is jogged down eastward by a fault and the shales that underlie it conformably probably extend to the creek bottom. There is one fault of 40 ft throw at the bar.

24.82 = top of red sandstone at cross bar; at the creek below 25.03.

There are many slides from the Tertiary walls of the cañon and the cross bar is 'if itiun.'

The Tertiary section is shown in Figure 1.9.
Just above the cross bar fault I can see level beds washed by the creek.

The most distant point of the red cap that I can see is W from me and its angle is –1¾ᵃ ᵃ.

{p. 11}   **Tuesday, June 29.** Last night at Salina again.

Up the southern side cañon on the line of fault. The Tertiary limestone (= 1 on Fig. 1.9) curves down with a max. dip of 45° and is covered by red clays and white sandstones which confirms the last idea with regards to the north fault-cañon (p. 7). The red series must show a depth of 7–800 feet. It is unevenly eroded and upon it and the limestone rest gray volc. sand covered by trachyte.

Found *Inoceramus problimatica* and *Oohaia* [= *Oonia?*] in 38 [Fig. 1.7] firming the presence of Cretaceous.
Wednesday, June 30, 1875
Salina 25.93 (8.25 am)
    25.90 (.12 n)
Salina Hogback

{p. 12}   At this point the beds inferior to the limestone are completely covered by its debris but farther north the point homologous with 7 is revealed as composed of green gray calc. shale

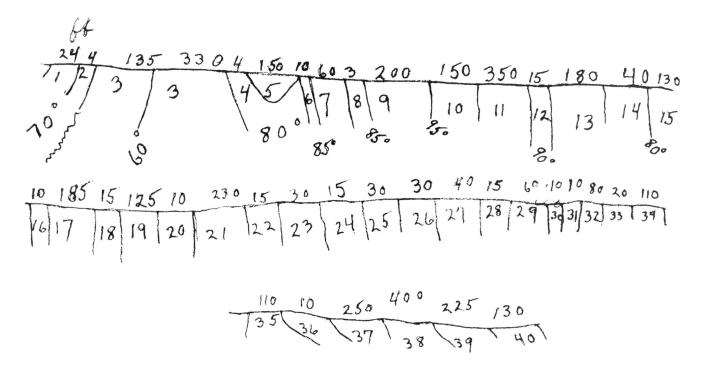

Figure 1.7. Detail of the section measured along the north side of Salina Canyon. 1. red shale; 2. white soft sandstone (strike NNE mag); 3. red shale partly concealed. Strike N; 4. red sandstone; 5. cut out for 150 ft; 6. red sectile sandstone (strike N0; 7. red shale; 8 coarse red sandstone; 9. red shale; 10. red shale and red sandstone (the beds I visited yesterday); 11. red shale; 12. coarse sandstone, brown; 13. yellow 'aren' shale; 14. yellow sandstone soft; 15. red and yellow 'aren' shale; 16. coarse conglomerate with pebbles of limestone, sandstone, and chert, calc. cem.; 17. red shale; 18. sandstone and conglomerate, gray; 19. soft, unseen; 20. gray sand and conglom.; 21. unseen; 22. yellow coarse sandstone; 23. red shale; 24. gray sand and conglom.; 25. red shale; 26. coarse sandstone, yell. strike N[m]; 27. unseen softer; 28. conglom. and yell. sand; 29. unseen, softer; 30. sandstone, red; 31. coarse quartzite conglom.; 32. yellowish sandstone; 33. coarse gray conglom.; 34. yellow sandstone and some red shale; 35. unseen; 36. yellow sandstone; 37. red shale mostly covered; 38. yellow sandstone with some shale, and a calcareous sandstone at top. 'I prit'; 39. red shale poorly seen; 40. massive yellow sandstone.

Figure 1.8. Section in Figure 1.7 continued. 41. yellow shale with sandstone at top; 42. yellow and red shale and 'bituminous' shale; 43. shales yellow, gray and perhaps red not well exposed. They are masked by red earth formation and by blocks of the surface hard sandstone which has here increased to a depth of 100 feet. In 43 there are some bituminous streaks and probably coal; 44. massive yellow sandstone, strike 195[a a]; 45. yellow shale with some sandstone; 46. Unseen to great cross bar, dividing the valley. S of the creek however a few sandstone beds out crop and show the dip to be 25°.

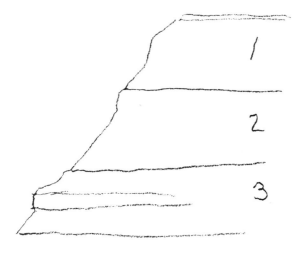

Figure 1.9. The Tertiary section. 1. cream lime; 2. green gray; 3. red.

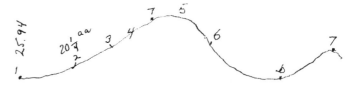

Figure 1.10. Profile and section Salina Hogback.

an at 1 – 25.94	t from 1 = 20¼^a ^{ao} ('fward')
an at 2 – 25.82	1 from 2 = 14½^a ^{a} ^{o}
an at 3 – 25.775	1 from 3 = 16½°
an at 4 – 25.70	2 from 3 = 25°
an at T – 25.62	T from 3 = 29°
an at 5 – 25.59	T from 4 = 31½°
an at 6 – 25.60	1 from 4 = 17¾

dip at 2 = 52°	cherty sectile cream limestone
dip at 3 = 49°	strike 195^a ^{a} from 2 to 4
dip at 4 = 45°	dirty massive cream limestone
	'fm top of 6'

an at b = 25.51	b from 5 = 22½°
an at 7 = 25.72	6 from 5 = 27°
	T from 5 = 21°
	b from 7 = 19°

resting on [*something omitted in notebook*]. The thickness of red cannot be made out and its dip is hard to discover and it cannot be distinguished from the non-conforming Jura. The green gray is about 500 ft. Highest point of hogback 25.18.

July 1st Thursday. Salina to camp 2 in Salina Cañon, near coal opening.

Friday, July 2, 1875.

Camp 2 in Salina Creek 20 ft above creek. 25.53 [5260]

Bot. of red cap	25.48
top of red cap	25.41

Red cap here is shale and some sandstone.

Slides to 25.70 then the gray green in place. It probably underlies the slides. The green-

{p. 13}   gray is calcareous shale weathering in a soft slope for the most part but one belt is hard enough to hold a cliff in much of the escarpment.

This begins at 24.97 and ends at 24.90.

From this point I travelled on slides of the cream lime to 24.81 where I find it in place. Looking across the creek NW I can see the c.l. [*cream lime*] and the hard band of the g.g. [*green gray*] distinct with 50 ft of g.g. shale between but NE they merge in one yellow cliff. Hilltop 24.55

Looking SSW across 'Soldiers' Cañon (Fig. 1.11). The red beds above the c.l. seem 800 ft thick and the tabular trachyte appears newer than the fault. To my surprise "black cap" seems as high as the trachyte table.

Changed position ½ m E. 2 m beyond is a N-S fault with E throw and beyond the beds are inclines at far greater angle than here. The dip here measured W is 5°; SW 8°.

{p. 14}   After dinner we started up the cañon. At the confluence of the Gunnison fork of the creek [*Gunnison Fork = Salina*

Figure 1.11. View across 'Soldiers' Cañon.

Cr. north of Salina Canyon?] we cross a great fault. The least rock visible is a shale series of 250 feet beneath the Red Cap—red and gray shale with yellow sandstone fillets. There comes in on about the level of the R. Cap what seems to be a limestone (2 in Fig. 1.12) covering shale. This shows a little to the N but not the S. On the immediate 'shore' on both sides are 50 ft. of soft yellow sandstone overlying as much purple shale. I do not connect this definitely with 2 and suspect the fault is divided. Trachyte debris to this point but not of local derivation. It seems to have come down Gunnison fork when the flood-plains were 75 or 100 ft higher. Immediately beyond the fault the dip does not increase but in a half mile or mile it does. Looking at G fork I measure it at 5°.

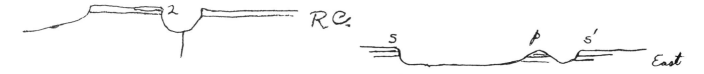

Figure 1.12. Possible fault in Salina Canyon.

Figure 1.14. Cross section at Puzzle Hill, p. 16. The hill is horizontally bedded and believed to be faulted down between older beds at S and S′.

No. I am wrong. There is no fault. The red cap is the bed 2 changed to yellow. Below it are red, purple, and gray shales, say 600 ft. The soft yellow

{p. 15}   calcareous sandstones on which I sit and which I can see further up stream to have a depth of several hundred feet. The dip here is SW and 5°. Farther north it seems to be 10°.

The sandstone continues all the way to camp 3 which may be 1,000 feet below its summit.

Dip 10°W. with a dip also N and S from the cañon.

Saturday, July 3. Camp 3 on Salina Creek to camp 4.

As we start we find shale intercalated with the sand and coal seams. (There was some coal last night.) Then comes more sandstone making a total thickness of probably 2,000 ft. The only fossils in all this are a few pinnate leaves (Fig. 1.13), *"Taxodium"* perhaps but broader. Coming into the valley which runs N and S at the head of the cañon. I see ahead Howell's puzzle hill of shale with a sandstone bluff behind it.

Dinner and 'trout' and rattlesnake on Salina Creek.

Figure 1.13. Pinnate leaves in coal-bearing beds along Salina Creek.

{p. 16}   After dinner I ride over to Howell's puzzle. The hill is made up of variegated (red, white, 'lead' and c [*cream?*]) shales with fillets of white limestone. In the limestone *"Physa."*

[*Fig. 1.14 is cross section at Puzzle Hill.*] s and s′ are in a line about 2 m apart and they are 'straight' cliffs with a SSE trend. I regard them as lines of fault and the puzzle as a superior bed, the summit of a lump that has fallen down thousands of feet. The structure must end in 5 or 6 miles southward. Top of p 23.82. A hill a mile north is 100 ft higher.

Northward the structure trends toward or east of Mooseniah [*Musinia Peak*]. S. rises to the north and S′ falls. There is most limestone in the cap of p.

Near the east base of p is a bed of conglomerate or cream sand (24.05).

The same 'siruk' mass appears

{p. 17}   south of Salina Creeks (P′) where it is covered by trachyte debris from the south.

From camp 4 it appears that S′ is coarser than a mass S″ immediately east of it and dips N at a greater angle. S″ is probably a little higher than S. The masses are separated by 'shear' faults with no signs of flexing.

See sketch on p. 18 (*Fig. 1.15*). The Musinia Plateau is built of strata superior to S″ and resting on 'them'; it is one mass with S″. Musinia Peak may be one with S.

Assuming that P is Tertiary superior to Musinia and that S, S′, and S″ are Cretaceous sandstone then the faults by which P has subsided are 5–6000 feet. The fault x trends 340ᵃ ᵃ. The fault y trends more NW and probably joins z at the south (Fig. 1.16).

{p. 19}   **Sunday, July 4, 1875.** Camp 4. 23.88 (7090) [*Camp 4 on Meadow Creek just south of junction with the south fork of Salina Creek. See Fig. 1.15.*]
Eastward 'smuk' some Tertiary at same height.
In Meadow Cañon 23.82; Divide 23.34; Gilson's coal prospect 23.66.

From camp SE to Meadow Cañon are Ter and Trachyte. Where we strike the cañon it parts the Ter. and Cret. but soon it passes into the Cret. sand ('Sabel'). Descending eastward in a gulch tributary to Ivie Cr. The coal we see is in several bands netting "7 ft".

D E u l i e n 10 May 1832 [*I find no explanation of this entry.*]

Hilltop Sta. 1 (Graves) 22.76

The sandstone we have been travelling in dips W at 1–2°. It is capped here by trachyte. [*This probably refers to one of the lava capped benches of Cretaceous formations extending north from Mt. Hilgard, but Gilbert sometimes uses this phrasing to refer to trachytic debris.*] Northward the sand makes an E facing cliff from which rise the sources of the San Rafael. The profile of the cliff (10 m.N.) is shown in Figure 1.17. Then eastward the country falls in terraces of west dipping beds. Just below Cliff one there seems a low anticlinal, exhibited in outlying islands and 20 miles N is a mass several miles

{p. 20}   size slidden (an 'Clagan') from Cliff 1.

Between Cliff 1 and Cliff 2 is a smooth 'looking' valley crossed at right angles by east flowing creeks. It too has its buttes showing it to be a cliff of erosion. Its profile is shown in Figure 1.18. The plain east of it shows much red.

Figure 1.15. View of Musinia Peak and the Plateau from Camp 4.

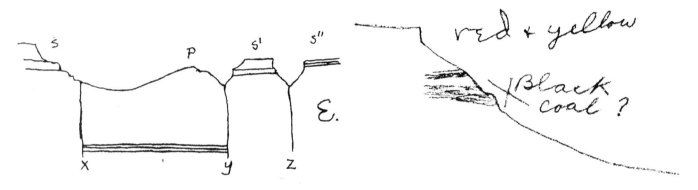

Figure 1.16. Cross section showing faults at Puzzle Hill (p).

Figure 1.18. Profile of Cliff 2.

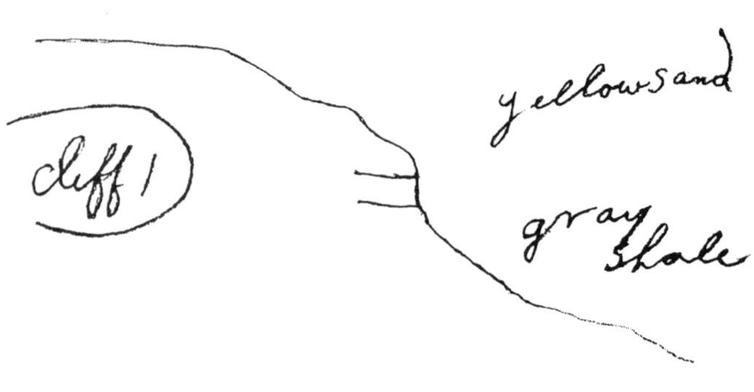

Figure 1.17. Profile of the bluffs at east side of the Wasatch Plateau as seen by Gilbert looking north from near the head of Ivie Creek.

In the 2nd cliff I can detect no displacements and I can see it for 40 mi. N.—not at all south of this point.

Cliff 3 shows a greater dip, a swell to the ENE, and a very decided fold of some sort due east (toward Elk Mt. [*not identified*]). From EE to SE the structure of the foreground, i.e., the space from cliff 1 to 2, is masked by trachyte debris which retards erosion of the sandstone and breaks up Cliff one into uneven foothills. These have their maximum 10 m away ESE or toward Henry Mts. At the gap in it E. of here the 2nd Cliff turns from N and S to NW to SE.

From SE to W (via S) the cuestas overlook this and are trachytic close by are 'in' outcrops of the sandstone (Howell's No. 4) all dipping westward at angles less than 5°. Just N of the Ridge is a great red escarpment and

{p. 22}   the same shows at all points to the N and NW of that outcrop. I take it to be the Tertiary of the Puzzle but a great deal higher.

Looking back (N 55°W) to camp and to Musinia and Musi-

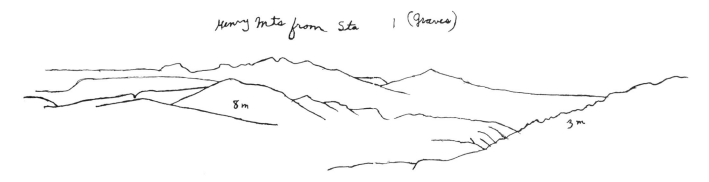

Figure 1.19. View to the Henry Mountains from the hills south of Ivie Creek. Distance about 50 miles. This sketch is not referred to in the field notes.

nia Plateau I can see sandstone that I take to be conformably superior to that of Cliff 1. [*Gilbert may have been looking at the Flagstaff Limestone.*] It has a dip greater by several degrees than that of Cliff 1 and the increase is gradual. I cannot see those 'Catlens' and stone run into M. Plateau and it is possible that it is a duplication by faulting.

Fork of Ivie Creek, 23.70.

July 5. Camp 23.925 (7060)

The trend of the Meadow Cañon fault is NW. Foot of hill 23.50; top of hill 23.22; Diff. –300 ft.

We climb a hill near the Basin divide we crossed yesterday. There is no fault along here. The dip (SW) is 2°.

Ahead towards Mt. Hilgard '   ' we can see enough outcrops to warrant belief that everything is conformable and unfaulted to the pink rocks [*Tertiary*] ahead.

Half mile ahead, on same horizon as hill 'oF' 23.225.

{p. 23}   This is 500 ft higher than Cliff 1.

To 23.60 sandstone continues but there is more shale with it and at 22.50 the shale gets reddish. This must be 2 miles from the last note (p. 22).

At Bell's Sta 12   22.35
                              32

We strike just above the reddish shales a white marl with some limestone. The first sandstone below (say 22.37) may be the sandstone at base of Howell's purple. All is still conformable and unfaulted but the dip is locally less or even reversed. It is resumed immediately. At about this point we pass from Tertiary to trachyte. Under the trachyte of Hilgard, 'Gilsen' and 'Emery' ridge, is Tertiary. To the east of this point is No. 4 and its banding running S toward Hilgard for 3 miles then turns E and afterward S again.

Dinner Sta 22.30 (89.76). Divide 'E' [*over a W?*] of Hilgard 22.075

Camp 5.

Climbing a hill eastward before supper we have a fine birds-

eye view of a moraine in another branch

{p. 25}   the main branch of Moraine Creek valley [*Head of Sheep Valley?*]. The main moraine is terminal but across the valley is a short lateral. The terminal bears a number of lakes.

Tuesday, July 6. Camp 5   22.06   ?

Climbing Hilgard.

The Blade [*Mt. Marvine?*] and Craggy head [*Mt. Terrill?*] are connected by a ridge of white rock horizontally bedded and a few hundred feet lower pink rock shows.

Thousand Lake Mt. in like manner shows white and pink (Fig. 1.21). The lava forming its table is not deep (3–400 ft?). That and Aquarius beyond are flat topped, plainly due to broad flows and are in contrast to the ridges in this neighborhood. Are the ridges dikes or are they the filling of pretrachytic valleys? The broad plain west of T Lake and Aquarius is all dark colored and graded by drainage lines to the DD [*Dirty Devil River*]. My guess is that

{p. 26}   it is floored by trachyte debris.

Summit Hilgard   20.38   (11,420)

After an hour   20.38

The lower end of the moraine of Moraine Creek ¼ mile south of a line joining Hilgard and the Blade. Thence northward the valley is morainal for two miles abounding in ponds.

The plateau west of Fish Lake has a slope arching from S to SE and is in perfect accord with the mt. E of the lake. Has the bottom of that valley too fallen out? [*Fig. 1.D*]

Under the E face of 'Snowy' ridge is a pond. Snowy Ridge, Craggy Head, and a ridge between them are monoclinal trachyte masses dipping SW 10–15°. The Blade I do not make out. It is improbable that the Tertiary ridge between was covered by lava.

These west dipping masses are probably in sympathy with the Musinia fold system and separated by a set of faults. Divide Valley is masked by slides and trachyte debris but it is not explained without a fault and a double fault meets its case most easily. In that case it is a prolongation of Howell's Hole.

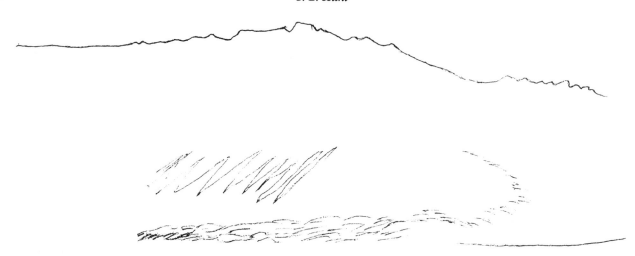

Figure 1.20. Unlabeled sketch on p. 22. Possibly The Blade.

Now I can see the companion drop to that of Howell's Hole, on the north fork of Salina Creek. It has a less southward extent seeming to stop at Salina Creek. The creek runs east of it and above

{p. 27}   perhaps crosses it.

Nebo has a short crest. Musinia Plateau is of great extent and of harmonious dip except part that rises higher at the north suggesting an E-W fault and a little knob over against Musinia. The general dip of the plateau is E and less than 1°. Where the creek east of Divide Valley joins this mt—half a mile from here and 1500 feet down are two ponds—slide-born.

At the northeast base of Hilgard there must be Tertiary under trachyte debris and beyond is No. 4 under trachyte debris in part.

The tabular top of Hilgard is not over 1/3 mile broad, a center at N in contour and sloping ESE at 7°. In a mile it changes to two ridges walling a valley which runs south two miles and then turns to SE. A mile from its origin the E ridge divides sending a spur ESE and several miles long.

What I called the 3rd Cliff two days ago now looks like an anticlinal swell with a longer diameter N and S. It may be the one Howell saw from 1000 Lake.

Eastward, toward S. tangent of S. La Sal, there appears first cedar (trachyte drift) foothills, a monoclinal fold with NW trend. I cannot trace it far (dip NE).

{p. 28}   Its chief cliff is red and back (NE) of that are white beds.

Toward Abajo and N. Henry the red plain is dotted by some dark topped tables [*areas of diabase sills at south end of San Rafael Swell; see Notebook 10, Figs. 10.15–10.17*] most of which are level but the nearest dips SW toward the main mass of Henry. Red sided tables look like a broad synclinal with SE trend. The

Figure 1.21. Thousand Lake Mountain as sketched by Gilbert from Mt. Hilgard. p = pink; w w w = white.

SW rise stretches toward 1000 Lake and is exhibited by a line of white cliffs which starting SE curve to S and pass between Henry and Aquarius.

(See p. 3 and foot.)
2100 ft above camp.

Wed., July 7. Camp 5.   2293   ?
Looking down on the Moraine I read 22.85. The foot of the glacier was 300 ft lower. Its length could not have been more than 3 or 4 miles. It had two heads lying on opposite sides of the ridge between Snowy and Craggy. For two miles or more the eastern valley is morainal. Many of the ponds are swamps. Others are beaver ponds as we find in crossing. It is hard passing over it and the road was necessarily

{p. 29}   sinuous. The heterogeneous pellmell heaping is characteristic and corresponds to same morainal surface at the east. Climbing to the divide westward we find at top 200+ feet of white, chalky, Pink Cliff limestone, level bedded and with level

Figure 1.D. View northwest across north end of Fish Lake Valley showing glaciated valleys cut into block faulted lavas. This is a fault valley, a graben, in the mid-Tertiary lavas, as Gilbert discovered, p. 1–30, 31.

crests forming the divide for a mile. Trachyte is N and south of it and the relation not seen. This Tertiary was not covered by trachyte.

Looking into Summit valley I see no moraines unless some small low-placed laterals.

Aneroid on Divide	21.10
Aneroid at first	
Observation Summit valley	22.10
Foot of Summit valley	22.40
Camp 6	22.88

In Summit Valley there was no general glacier, but some short lateral valleys probably sent glaciers down into Summit Valley. All these heaps are on the west side. Fish Lake valley is similar. All its tributary arroyos are on the west, and the moraines (at least at the north) pertain to them and not to the main valley.

It still looks as though Fish Lake valley has

{p. 30}   dropped out. The sides are markedly straight and steep and the lateral arroyos are incidental and do not control the contours (Fig. 1.22).

Thurs., July 8. Camp 6 - Fish Lake   22.43   [8830]
Marsh   22.47   swale beyond   22.55
Crag near hilltop   21.79

Dip and 'E' to foot of lake 1°10'. What was the origin of Fish Lake valley?

Figure 1.22. Plan of valley wall and arroyos at Fish Lake.

Figure 1.23. Cross profile of Fish Lake.

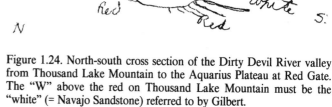

Figure 1.24. North-south cross section of the Dirty Devil River valley from Thousand Lake Mountain to the Aquarius Plateau at Red Gate. The "W" above the red on Thousand Lake Mountain must be the "white" (= Navajo Sandstone) referred to by Gilbert.

1. Its walls are of trachyte. Are they massive eruptions over dikes—hills of eruption? No, for they exhibit in them as components a bedded structure and their slopes are too steep and angled at top. A profile on the east side measured 33°. Figure 1.23 is the cross profile.

2. Is it the result of aqueous erosion? No, for it is too broad to have been cut from the trachyte since the age of the trachyte and its side cañons are disproportionately small. Moreover it is now the summit valley with but a 30 ft bar to prevent a reversal of its drainage direction.

3. It is not the result of glacial

{p. 31}   erosion for the glaciation of this region is but slight—entirely inadequate to the making and shaping of a great valley.

4. It remains to suppose that since the trachyte epoch the floor of the valley has sunk (or the adjacent hills have risen) an amount equal to the present depth of the valley plus the depth of its detrital filling, plus the loss of the uplands by waste—an amount somewhere between 2000 and 3000 feet—say an amount very nearly equal to the elevation of Fish Lake Mt. above camp.

Again, what was character of the glaciation of the valley? Was it filled with ice? I suspect not. At the north end, the valley is divided by a large trachyte mass slidden apparently from the west cliff. It is preglacial and between it and the west cliff arose a glacier, a lateral moraine of which is cut by the outlet of the lake. A cañon a mile or two further south threw out a deep moraine nearly to the middle of the valley, and a low bar which stretched from it and nearly divides the lake is probably morainal. The moraines of Summit Valley seem to belong not to this great valley but to western side cañons and amphitheaters. The bar at the head of the lake I suspect to be trachyte in place and must examine as I descend.

From my perch I can

{p. 32}   look [*west*] across Grass Valley and the question arises whether that too is not due to two faults instead of one. From the mountains near Marysvale, Monroe and Glencove there are grand slopes converging toward Rabbit Valley and abruptly interrupted by Grass Valley just as the slope from Fish Lake Mt. to Rabbit Valley is interrupted by Fish Lake valley. Further south the west wall of Grass Valley has no tabular character and its sculpture is alpine.

Crossing to the east side of the ridge I look into Rabbit Valley. It is a grand center of drainage with no visible structures—all trachyte.

The pass between Thousand Lake and Aquarius (Fig. 1.24) seems a monoclinal with southward throw. [*This view is about along the axis of the Teasdale anticline.*]

Rain.

Descending to the bar at the head of the lake I find that it includes a south facing bluff of trachyte in place 30 or 40 rods long. So it certainly is not a moraine; nor is there any moraine at the head of the lake.

{p. 33}   **Friday, July 9, 1875.** From Camp 6 in Fish Lake to Camp 7 in Rabbit Valley.

Descending the lava slope of the Rabbit Valley fantail we find it to be trachyte in place and not drift so there is no chance the sedimentaries beneath it will be seen. If M. L. [*refers to Thousand Lake Mtn. M = Roman 1,000*] is not separated by a fault then its lava is very ancient by contrast—an hypothesis not borne out by the general aspect of the two fields. The whole visible (west) face of M. L. shows level strata but there is red at the right on the same level with white at the left and middle. This suggests but does not demonstrate a fault. The fold in the gap between M. L. and Aquarius seems just as yesterday and now exhibits a trend ENE[mag].

A large outlier N of M. L. may dip to the NW.

This Fantail lava has a gray, sparsely vesicular part with crystals of sanidine only. Its variation is chiefly in paste.

There have been local small faults in the Fantail lava.

Where we strike Rabbit Valley there is a valley ½ mile broad trending NE and at right angles to the drainage which is by cañon.

In this

{p. 34}   valley I find on the NW side a marly deposit with *"Planorbus"* and *'cydas'* and on the SE, evidently underlying the lava are red and pale green shales. They contain gypsum and 'may' dip to the N.

Climbing the hill E of camp I find it narrow and the cañon we are to follow tomorrow is short. The valley behind, W of the ridge, curves around to the SE and E and has yet another outlet further east. The lavas in the ridge are exposed in the cañon. They dip toward Rabbit Valley and a breccia lies below a compact trachyte.

Sat., July 10. Camp 7   Rabbit Valley   23.93   (7060)
Head of DD Cañon   24.17   [*at Red Gate*]

The first rock I meet at the river cañon is the Shinarump conglomerate which is here a yellow sandstone. It has below it about 200 feet of red shale and caps a bold escarpment north of the river for 3 or 4 miles to the east. For most of the way it is level on the north but at the head of the cañon it bends down to the waters edge.

Section from DD R. toward Aquarius

{p. 35}   River at about the base of the Shinarump 24.14   (6820)

2)   Yellow sandstone 60 ft dipping 8°. Trend of valley N 75°E offset to   24.07 (6870)
3)   Chocolate (and gray) shale   100 ft
4)   Chocolate sand   3 ft
3a)   Chocolate shale   30 ft
4a)   Chocolate sand   3 ft
5)   Variegated clays purple, red, gray (fossil wood)   300 ft
6)   yellow and brown soft sandstone   20 ft
        24.01   [6970]
        2 from 6—4 1/2$^{d\,a}$, dip 10°, strike WNW
7)   Green and red shale   60 ft 2390 (7030)
8)   Massive   lam sand
        Top 6 from top 7   18da
8a)   red   300 ft
8b)   yellow banded 'untried'   900 ft
        Sta X is 2/3 'up'   8b dip 25$^{a\,a}$ 23.72   (7300)
        6 from Sta x 28$^{d\,a}$
        2 from Sta x 12 1/2$^{d\,a}$
        Offset to 'Satun' which [*is?*] 75 ft higher in 8b and near its summit.
        Sta y   23.62   (7410)

8c)   is red
8d)   is white

From Sta y it looks as though the western foot of M. L. included a mass much lower than the main mt. lying west of a N-S fault. The red Trias escarpment terminates suddenly and is succeeded at the same level by a slope of volcanic debris through which pale yellow or white beds show occasionally.

The most distant point I can

{p. 36}   see on this fold is 8 miles away ESE = 104$^{a\,a}$. A white sand is there weathered into an eolian cliff pinnacle.

ENE there is another displacement evidenced 10 mi away but I cannot make out its character save that there is a northerly dip. Neither can I tell why the Shinarump escarpment across the river suddenly ends. The vermillion cliff opposite holds 3 or 4 mi [and] is then recessed by a cañon from M. L. and beyond exhibits a gentle eastern dip. The Shinarump cliff is so high that it seems as though the Carb. must be nearly reached by the river.

Section resumed

Top of 8 c   (23.44)   Dip 30°   Sta y   8 1/2$^{d\,a}$
Top of 8 d   23.23   = 7880   Dip 20°   Sta y   14$^{d\,a}$
Top of 8 d   23$^{d\,a}$
9)   red sectile sandstone   15 ft
10)   yellow green sand and lime with fossils 30 ft
11)   yellow gray shale, soft with much gyps and 'one' band of foss. lime 200+ ft.

A later view explains the view NE. The Shinarump is continuous and so is the vermillion. The acquired eastward dip of the beds carries the Shinarump out of sight in 8 miles but it is brought up again by a fold. Where the beds become level south of the fold the sandstone "14" is on a level with the 25 of the Shinarump across the river.

(over)

{p. 37}   **Sunday, July 11.** Supply camp   24.04 [*see Table 0.1*]

The section of yesterday can be renewed stratigraphically up Supply Creek—

10)   yellow shale, sandstone, and limestone   30 ft
11)   yellow shale with band of red   40 ft
12)   gypsum [white]   10 ft
13)   yellow shale and sand   40 ft
14)   yellow sectile sand 'qtz'   30 ft

A later better observation makes base of 13 south of the fold lie on a level with the base of 5 on the north. Getting a better view in Supply Cr. cañon I see first that in the cañon there is a rapid dip westward that brings the highest beds visible in the hill to the east down to the drainage lines. The first dip is probably a part of the west Aquarius fold or fault [= *Thousand Lake fault*] + 600 ft + is the displacement exhibited.

15)   a series of red shales and gypsum beds in equal shares   130 ft
        top of 15 - 23.61 = 7410

base of 15 - 23.73 = 7280

The arroyo next west of this is the limit of sed. exposure. East of that is trachyte.

Monday, July 12. Suppy camp

The plotting of the DD fold gives the Trias sandstone 1700 ft and the shales down to the Shinarump 400 ft. Displacement 2200 ft.

In passing down Rabbit Valley two days ago we noted that the river at one point leaves the open and obvious way and cuts through a portion of the trachyte slope on the south—horseshoe-cañon-wise. [*Location just west = upstream from the Bicknell Bottoms.*] The river course may be either antecedent or superimposed—more probably the former [*probably both = anteposed*].

The slope of Hilgard and its summit illustrates typically the weathering of the trachyte. There is a transition from solid massive trachyte to the same seamed and loosened and to the same in boulders, rounded not by

{p. 39}   transportation and attrition but by waste. A hill capped by trachyte boulders may be the remnant of trachytic drift or of a trachyte sheet in place.

Camp 24.05

Tuesday, July 13. Camp 24.05

The shale below the Shin. [= *Moenkopi Formation*] is red-chocolate in color and filleted with pale shale tints which are made in part by shale but chiefly by gypsum. The chocolate is permeated by gypsum.

River 4 mi E of Supply Camp   24.19   (6770)
1)   massive red sandstone   15 ft
2)   red sandstone and shale. The sand rippled. More sand at base; more shale at top   60 ft   24.08 = 6870

The opposite (S shore) shows several drift terraces built chiefly of trachyte boulders. Their heights are: 30 ft, 100 ft (measured), 200 ft and perhaps another at 250 ft. Their heights increase westward more rapidly than the all[*uvium*] of the present floodplain [*Pleistocene tilting*]. The 200 ft makes a pretty island table 2 miles east of here (p. 57). Opposite here the eroded rock in the fold is the Trias and the first cliff beyond is Jurassic with 200

{p. 40}   ft of sandstone foundation.

3) chocolate shale with gypsum forming a bold palisade under the Shinarump.   400 ft.

There is a gentle dip N which anticlinals with the DD fold. There should be noted a tendency to fault exhibited in the sandstone. '8c' and '8d' are divided by a vertical instead of a 30° line. Still I am not disposed to think the purity of the fold is vitiated nor the measurement of thicknesses are notably wrong. Here (E end of palisades) however the appearance of fault is more decided; still it cannot be affirmed without going to the spot which is 3 mi southward.

ENE and 5 mi distant is a conspicuous fold tilting the Vermillion and Shinarump. Its dip is NW and its course NE or NNE. It meets the DD fold at or near the white [*Cockscomb*] crag. The

anticlinal of the two very likely brings up the Carb. The DD river probably cuts this anticlinal.

4)   yellow sand ('shin)   100   23.81 = 7190
5)   variegated shales with some sand; the harder beds chocolate (much wood)   300 ft   23.57 = 7450

{p. 41}   The great sandstone. Red or white. A fine exhibition of the accidental nature of the red. The white which forms the summit of the cliff comes down at one place to the very base cutting the red out completely. This is the explanation of the arrangement of white and red on the W. face of M. L. and perhaps in part of their distribution on the N face.

Dinner camp   23.51   75.40

The outlier NE of Aquarius [*Lion Mountain?*] is White Mesa Sand. and is about on a level with this. The meaning of that is that the two folds cross making an anticlinal E of the White Crag and a syncline west.

At   22.70 = 8500 the red changes to white
At   22.20 = 9090 the top of the great sandstone (6).
Gyps at 21.80

At 21.40 = 10,100 limestone (white like pink cliff) with black pebbles 'and brecciated' (Tertiary ?). The slope with these exceptions is covered by debris of trachyte from the summit.

As we climbed there has come into view a great monoclinal dipping NE and trending SE. [*Capitol Reef*]

{p. 42}   making with the DD fold an anticlinal. It holds its course for 50 miles SE and the cliffs face it all the way to the foot of Henry Mts. How it joins or crosses the fold noted this morning, I do not know. Its great cliff is the great sandstone. The highest points are white [*Navajo Sandstone*]. The Trias we have intersected today dips gently to the N. So the thicknesses are greater than indicated by the aneroid.

Camp 9.   21.07 = 105.20

SE on a spur.

The Howell fold (for the sake of a name) [= *Teasdale anticline*] continues NW as though to cut the NE base of M. L. Its throw is NE and in that direction from here it makes a synclinal with the great quaquaversal seen from Hilgard. The rock that is preserved in this synclinal is Upper Jura (?) and its escarpment is the Belted Cliff. The cliff is very persistent (Fig. 1.25).

Wed., July 14

Camp 9   20.98 = 10,630
Summit M.L.   20.50 = 11,280
M.L. that looked so

{p. 43}   level from below, undulates notably. From the Summit to the S edge the slope is 5°. The lava seems like that of the Dish—a gray matrix with feldspar and a decomposing mineral—hornblende?.

Going west to the edge of the Table I look down on the Dish and the shoulder. The shoulder is a grand exhibition of erosion by undermining. The long slope from the foot of the trachyte cliff to the first exposure of Jurassic is covered by ragged ridges grasslife

Figure 1.25. The Belted Cliff (= Morrison Formation).

and scantily timbered, of trachyte that has sunk by undermining and the work is not done. The nearest ridges contain large coherent blocks, the others only boulders.

'The Dish is structural.' Its rim is lowest towards Marysvale Mt., highest where it joins the Aquarius plateau. From the latter point a post-trachyte fault [*Thousand Lake fault*] forms the eastern boundary and separates it from the Aquarius and M. L. This fault increases in amount from 0 at the south to 4,000 or 4,500 ft. at Rabbit Valley. Perhaps it diminishes northward. Opposite this point it (or a part of it) is three miles distant. Northward the limit of the Dish is indefinite and the structure is complicated by the system of displacements to which the Blade, Hilgard, and neighboring ridges and valleys belong.

The structural valley in which lay Camp 7 ends in a mile southward but north continues 10 or 15 miles

{p. 44}   and includes a part of the course of the D. D.

The nearest exposure is lower Jura. The foothills beyond are of pale yellow and red soft rocks and are probably upper Jura, the two being separated by a fault.

Travelling N I see on the W shoulder five ponds of which 2 are of potable size. The trend of the Camp 7 structural valley is SW or SSW.

The shoulder at the NW shows red shales and gray shales within 1500 and 100 ft of the summit.

'From the north end' I see one pond on the NE slope.

The crest runs to a farther edge and continues five miles at least and perhaps 10.

Castle Valley Cliff = No. 1 is lost in the trachyte debris foothills east of Hilgard. No. 2 is Cretaceous and forms the limit of those foothills on the SE. I see no more of it hitherward. It is crossed at the east salient of the foothills by a cross fold with northward throw and would be thrown far westward but for the trachyte cover. It probably is be 'celled'. At the same point the westward or NW dip is increased which looks as though the fold might be somewhat anticlinal. The next cliff below (No. 3), between No. 2 and the anticlinal, is red—almost surely Jurassic and is the equivalent of the Belted Cliff. In its

{p. 45}   southward course it is jogged 15 miles by the cross fold with NE throw and its southern course is crowded close to No. 2. I get no further evidence of that fold. The bearing of its salient, of its reentrant a little E of N. Nearer by (10 m) and still NE the same cliff surrounds a peninsula projected eastward from the base of this M. L. or rather from the base of the N spur of M. L. (Bearings are worthless on this magnetic rock.) Only a few miles separate this peninsula from the synclinal table which the Belted Cliff limits on the SW. The table is 2–3 m wide at the NW end increases. There is reason to suspect that toward S La Sal a fault parallel to its course divides it. If it does then it is one with the fold which intersects No. 2 and No. 3. In a general way the rocks rise from this synclinal to the San Rafael quaquaversal. The Howell fold runs full strength into the foot of this M. L. along its E side. Its throw must be 4,000 ft. The base of the Jura reaches the foothills east of the N end.

'From the E Salient near the N End' I can see Howell's synclinal on the Howell monocline. Near the M.L. it is gentle on the NE (Fig. 1.26)

Figure 1.26. Profile of fold seen from the east salient near the north end of Thousand Lake Mountain.

{p. 46}   but further along (12 mi. say) it is stronger (Fig. 1.27) and it runs out on the E side; toward the saddle of Henry Mts. the SW side of this synclinal is that which makes an anticlinal with the DD fold.

I do not recognize the great island mesa NW of Henry Mts. but suspect a Cret. cap [*Cainesville Mesas*]. The same bed flanks Henry Mt. on the W, but not on the N and appears to rise toward the Mtn.

The largest lakelet of all is on the east side near the middle. Near it the Jurassic outcrop strikes the Mt. and changes from a NW course to a S course. The Island mesas are the equivalent of No. 2, Castle Valley [*but = Cliff C of Henry Mountains, see Table 0.1*]. There seems to be a monoclinal with E throw at their west base—a monoclinal that may join the San Rafael fold at the north and Howell's fold at the south.

'From the S. End' it seems doubtful whether the DD fold

Figure 1.27. The fold (Fig. 1.26) becomes stronger southeastward.

Figure 1.E. View north across the Red Gate at the Thousand Lake fault. Fremont River, Gilbert's Dirty Devil River, flows to the right into the Gate in foreground. Lavas like those at Fish Lake Valley (Fig. 1.D) cap the Aquarius Plateau (behind the observer, see Fig. 1.A), and Thousand Lake Mountain (Gilbert's M.L., center skyline) have been uplifted about 4,000 feet along with the anticlinally folded Triassic red beds on the east side of the fault. This is one of the places where Gilbert was able to distinguish two episodes of structural displacement—an early episode of folding that we now call Laramide and a later, mid-Tertiary and younger, stage of faulting. In this view a late-Pleistocene terrace (center) is cut off by the fault which, as Gilbert noted, developed in small increments along with the erosion of the water gap (see Notebook 1, p. 58).

{p. 47}   becomes parallel to the Howell. It may intersect and join it E. of the E. salient of Aquarius.

Either DD river is older than the trachyte and cut down as the Dish fault [*Thousand Lake fault*] progressed; or the DD outlet was a reentrant angle in the original field (Fig. 1-E). The latter is the more plausible and it is supported by the fact that now the trachyte is less widely separated at the west than at the E of the gap. The little covered island of sedimentaries E of the M. L. may be an allied fact.

Descending we camp where we nooned yesterday. The prevailing dip of the cross lamination in the Vermilion is south.
Thurs., July 15. Camp 10.   23.16 = 7940
Top of Shinarump overlooking the DD river   23.36 = 7710

It still looks as though the DD fold might be partly a fault between Supply Camp and the White Crag. The Shinarump comes very close to the Jura.

The crest of the anticlinal E of the White Crag is on a level with this point and its horizon is about the same as the base of this cliff (No. 1, p. 39). The throw of the syn.-mono. is more than this however and 'seems'

{p. 48}   more than double it, for a cross section through the crest of the anticlinal would exhibit the Shinarump as low as the river level and still running down. Southward the Jura fossil lime is about on a level with me and its dip is slight. This gives the displacement 2200 ft.

A stone dropped from the cliff strikes in 5½ seconds.

The northward dip is 1½[d] [a]

The var. shales [*Chinle Formation*] have much fossil wood. S of camp under trachyte debris is a Trias or volcanic sand. This indicates that the lava is substantially in place though only in boulders and that the fault lies between

{p. 49}   these hills and the Jura. The lava is the same basaltoid that occurs across the river.
Friday, July 16, 1875.
From Supply Camp   23.79 = 7210 to Camp. 11   23.67 = 7350

For 8 miles from camp I followed monoclinals of the DD fold chiefly on the upper Shin. Shales. Saw much silicified wood, but saw neither branches, bark, nor leaves, nor have I ever seen

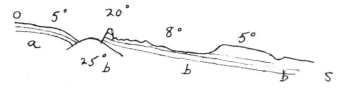

Figure 1.29. Profile and section at the White Crag. *a* is Red shale (= Moenkopi Formation) below the Shinarump. *b b b* is White Great Sandstone (= Navajo Sandstone).

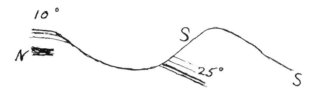

Figure 1.30. Section near camp 11, p. 50.

Figure 1.28. A mile of the Dirty Devil River under the Shinarump Cliff. This would be upstream from the anticline at Miners Mountain.

them that I can recall. If the tree be an exogen then branches would be indicated by knots, but there are no knots. Continued below.

The appearance that I saw yesterday of a fault replacing the DD folds in part is fallacious. What I took for a Shin. hogback is really Vermilion and the Shin. is locally thin and inconspicuous.

At a point halfway to the White Crag things are a little mixed. A local dip of 10° to the WNW is exhibited by beds I take to be Shin., and adjacent to them is the great sand dipping S and SE.

Climbing the hill I find it indeed Shin. but the dip is quite local. It *may* be a part of the syn-mono, but it looks more like a local anticlinal. Westward the beds have their normal position and eastward the cliffs do not retreat to

{p. 50} the south as they should if a monoclinal crosses here. Figure 1.29 is the way the White Crag looks from this point, looking ESE that is.

The Break of the synclinal is ENE and 15 miles away. The DD cuts the anticlinal on its N slope.

From Camp 11 I rode around a hill of Great Sand to examine the fold. Red shales and rippled sandstones [*Moenkopi Formation*] lie N of a little valley at N and White Massive [*Navajo*] sandstone south (Fig. 1.30). My first thought was that an abrupt dip intervened by which the Shin. and Var. [*variegated = Chinle Formation*] were passed under, but further examination shows that the N beds lie immediately beneath the S. It may be that the Shin. is disguised in thin bedded rippled sandstone but it is more probable that the Shin. and Var. are wanting. This will accord

Figure 1.31. "The Dirty Devil Gap [*Red Gate*] as seen from the White Crag." The view is about NW.

with my distant determination which made these red rippled sands the equivalent of that 300 ft.

{p. 51}   below the Shin. across the river, and further with the fact that the Shin. at my last station lies almost in juxtaposition with the Great Massive (G.M.). I see no fossil wood.

The inclination of the cross lamination is SE, instead of S as noted on M. L.

White Crag.   23.02 = 8120

The DD fold is marked by a valley 10 miles to the ESE but in the interval is concealed by debris from Aquarius. The course forward and back is straight, and the valley under us (N) is the counterpart of the 'rise' ahead and the gap of the DD.

The heavy synclinal is NE and 10 miles.

S of the valley ahead is the square [*a square, not spelled out*] top butte with level G.M. and Jura. N of it is gently inclined red shales and sand.

S from the Crag is G.M. carved out in valleys and before it runs under the trachyte debris, capped by a little Jura.

The San Rafael anticlinal is a very broad swell, perhaps half as broad as long. We seem in looking NNE to be looking at the end of it.

Looking NNE the rocks dip northward from the DD to the

{p. 52}   (Fig. 1.31). G.M. 5 mi. beyond. At the nearest point the dip is small but it increases eastward to 5° at a point 6 or 8 miles distant. From the cañon of the DD to the valley under square [*again, a square, not the word*] top butte is a plateau of lower Shinarump [*Moenkopi Formation*] that 'Watson' calls Round top plateau. Toward its edges the strata dip gently N, W, and S and its summit is broadly convex.

As seen from there the N side of the syn-mono achieves a max. dip of 8°.

Where we climbed M. L. the

{p. 53}   G. M. has a gentle N dip but immediately eastward it

dips S with the other side of the syn-mono. 5 miles S the same dip can be seen in the Shin. and still further it returns to the G. M.

The slopes of Henry Mts. are steep. The south tangent of Hillers is 30° for 2,000 ft. The N tangent of the N mass is even 50° but for a less distance—say 800 ft. [*Gilbert is probably seeing the north side of Bull Mountain at the NE corner of Mt. Ellen.*] The sculpture of the group is alpine and I expect to find them very old.

Camp = $11^{d\,a}$.

Through the gap of the DD fold valley is seen a butte [*a in the sketch, Fig. 1.32*] with reversed dip. It must belong to Howell's fold.

Wheeler's Crag
This White Crag (renamed *Ragged Edge*) is thoroughly seamed

{p. 54}   by uneven, interlacing, reticulated calcite veins, the greater of which dip northward 60° or about in unison with the steep north face. Only a few rods to the south the sandstone is free from them. They testify to the yielding under the shearing force (?). The same veins accompany the maximum dip in the sandstone near Supply Camp.

Ragged Edge 22.97 = 8170. At 23.15 = 7950 I am on a level with the top of the G.M. south of the DD fold. At the same height I am not greatly above the proper position of the 'base' of the G.M. north of the fold (if at all). The throw is about equal to the thickness of the G.M. if the Shinarump is absent. If the Shin. is concealed by fault, then the throw of the fold and fault is greater. **Sat., July 17.** Camp 11. 23.80 = 7200

Across trachytic debris over G.M. to Fold Valley (see Fig. 1.30 for section).

{p. 55}   The bearing of the DD fold is still ESE and I can trace it (from x, Fig. 1.33) 10 mi. I cannot trace it beyond Howell's fold but 8 miles away it makes a short anticlinal with that fold.

Figure 1.32. "Through the gap of the DD fold valley is seen a butte [*a in the sketch*] with reversed dip. It must belong to Howell's fold," p. 53 [○ *Top Plateau = Miners Mountain;* □ *Top Butte = Lions Mountain*].

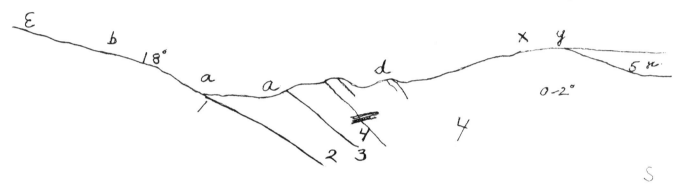

Figure 1.33. Section across fold valley.
1. red shale with red and yellow sandstone. It may belong 'just' below G.M. or Shin. 200 ft seen.
2. 'Terraces'
    b from a = dip = 13^{ea}. × from a = 12^{ea}.
    a = 23.65 = 7380
    c = 23.625 = 7400. a from c = 4da±
    x from c = 16¼^{ea}
3. = purple hard shale   20 ft seen
4. =G.M.   dip at d 30°
    Level of b = 23.45 = 7600
    Level of e = 23.25 = 7730
    b from x = 5da
    x = 23.07 = 8050
    'e' has lower rock than 'b' (75 ft?)
5. Red sandstone and shale
6. 20 ft above 5 is gray limestone with fossils

The Trias series is greatly changed and the absence of the Var. slaking shales and fossil wood is conspicuous.

The Round topped plateau has a 5° dip to the NW on its NW side which brings its top rock almost down to the creek I last crossed. To the SE it does not dip save as it comes within the influence of the DD and Howell folds.

From the SE corner of Square top [*square figure, not the word*] I can see that the Howell fold, beyond the junction of the DD has a gentle obverse dip so as to be slightly [*word missing?*] and the escarpment which shows it carries the G.M. in full force clear across the fold. At

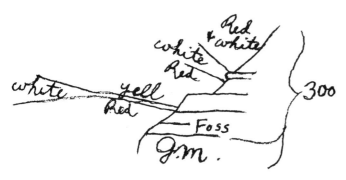

Figure 1.34. Section on p. 56; not clearly referred to in the notes.

Barometer readings at the top of the G.M. south of the DD fold—
At [square] Top   22.93 = 8220
    At White Crag   23.15 = 7950
    at the Gap   7600(?)
    It would appear that the rocks rise to the east slowly.

Note to p. 39. The point to explain about these terraces is how they have been produced without concomitant terraces in the Dish circling Rabbit Valley. Certainly if Rabbit Valley was a lake while the river below received these benches it could not fail to receive still more conspicuous benches and such benches would

{p. 58}   be more permanent on the gentle slopes of the Dish than on the steeper slopes of the river bank. Is it not probable that the fault west of Aquarius and M. L. has progressed 'pari passu' with the erosion of the river gap?

{p. 59}   Altitudes corresponding to Barometric inches - unreduced

Inches	Feet		
18	14,000		
	140	for each	1/10 inch
19	12,600		
	135	------	1/10 inch
20	11,250		
	130	------	1/10 inch
21	9,950		
	125	------	1/10 inch
22	8,700		
	120	------	1/10 inch
23	7,500		
	115	------	1/10 inch
24	6,350		
	115	------	1/10 inch
25	5,200		
	110	------	1/10 inch
26	4,100		

Cash items 1875
'Bob.' hammers and chisels June 7 $ 8.95

Cash on hand June 24   52.10
Lariat of rawhide J. 28   1.50
June 30. Pills   .20
July 1. Hotel in Salina   2.50

Letters list
June 23   ---   No. 20   GSG   (card)
June 25   No. 21.
June 27   No. 22.   GSG
June 29   No. 23.   IWP - EEH
July   1   No. 24   JJS., Mrs. Ellis
July   8   No. 25.   Home
July 11   No. 26.   Home
July 15   No. 27.   Dutton

{p. 56}   the junction of the two the G.M. becomes the floor of the upper country along the anticlinal and only the cañons reveal lower beds.

Still I see no sign of the DD fold beyond the Howell.

From the DD fold to the foot of the Aquarius the G.M. is dominant and level and a number of its buttes are capped, like this one, by Jura. Many surfaces are planed off like this and covered by trachyte debris—planed in long slopes descending along old levels of drainage from the Plateau. One that I can measure slopes 3½°. They are 500–1000 feet above present drainage grades.

It is now indubitable that the most northerly of the group of mesas faces at its western base a monoclinal fold with E dip and NNE trend. This curves about and joins the Howell, and north-ward trends toward the E flanks of the San Rafael arch.

The eastward dip of the G.M. east of M. L. and the west-ward dip of the O-topped plateau [Miners Mountain] conspire to run out the syn-mono so that it is a slight affair at the west where it hits the DD fold

A distant profile of the Jura shows the same white gypsum and red and yellow beds as at Supply Camp. [Perhaps this is illustrated by Fig. 1.34]

Summit of Square [a square] top butte. 22.31 = 8950
1) Trachyte debris over red shale to 22.50 = 8730
2) yellow green shale with some red shale and frequent bands of gypsum
3) a ten-foot bed of gyps at 22.67 = 8530

{p. 57}   4) red shale and gyps to 8450 = 27-74
5) pale yellow aren. shale changing at 8420 to
6) yellow splintery sandstone or limestone, fine grained 15 ft.
7) same as 5 to 8310 where is
8) a harder bed with fossils 5 ft.
9) a little below this is an outcrop of gypsum
10) red (above) and yellow (below) shale to 22.90 = 8150 where is a
11) thin red foss. lime ('campronestes')
12) 'unser'
At 22.93 = 8220 is G.M.
Camp 12 - 23.13 = 7970

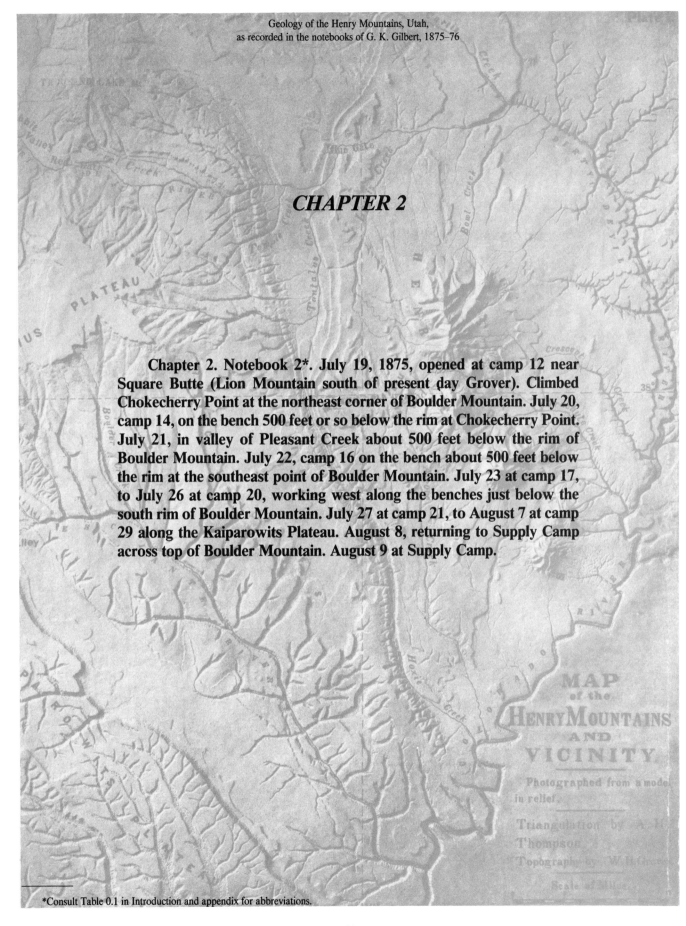

CHAPTER 2

Chapter 2. Notebook 2*. July 19, 1875, opened at camp 12 near Square Butte (Lion Mountain south of present day Grover). Climbed Chokecherry Point at the northeast corner of Boulder Mountain. July 20, camp 14, on the bench 500 feet or so below the rim at Chokecherry Point. July 21, in valley of Pleasant Creek about 500 feet below the rim of Boulder Mountain. July 22, camp 16 on the bench about 500 feet below the rim at the southeast point of Boulder Mountain. July 23 at camp 17, to July 26 at camp 20, working west along the benches just below the south rim of Boulder Mountain. July 27 at camp 21, to August 7 at camp 29 along the Kaiparowits Plateau. August 8, returning to Supply Camp across top of Boulder Mountain. August 9 at Supply Camp.

MAP
of the
HENRY MOUNTAINS
AND
VICINITY.

Photographed from a model
in relief.

Triangulation by A. H.
Thompson.
Topography by W. H. Graves.

Scale of Miles

*Consult Table 0.1 in Introduction and appendix for abbreviations.

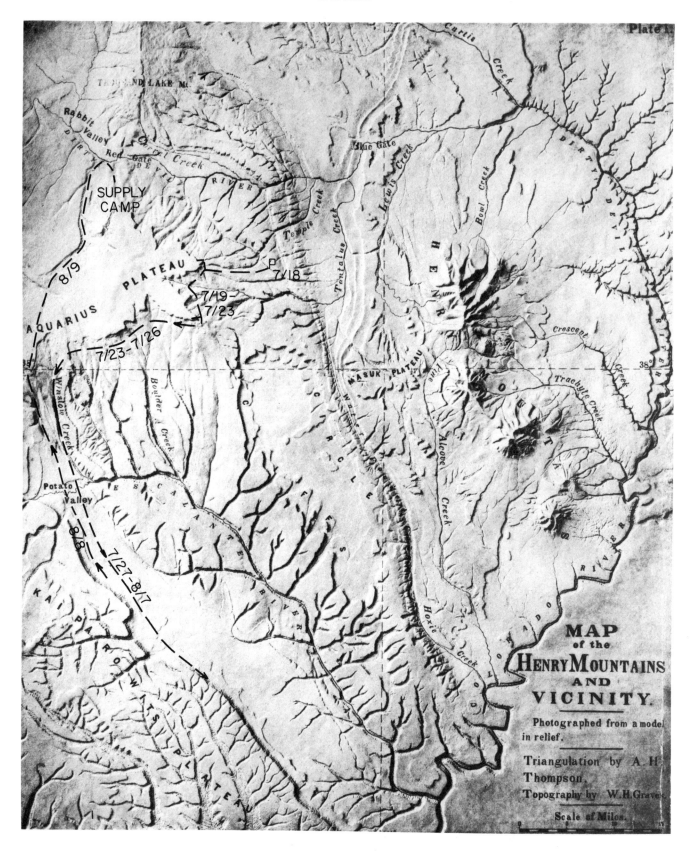

Figure 2.A. Landform map showing Gilbert's route around the east side of the Aquarius Plateau and in the Escalante River basin, July 13 to August 8, 1875.

Figure 2.B. View east across the Circle Cliffs uplift to the Henry Mountains. This view, from the present highway near Boulder, shows the west dipping Navajo Sandstone (Gilbert's G.M.) forming the west flank of the uplift. The numerous canyons incised into the sandstone are tributary to the Escalante River. On the skyline can be seen Mts. Ellen (L), Pennell, and Hillers (R). Gilbert reconnoitered this part of the Escalante River basin July 23 to 26, while Graves and the other topographers were occupying rim stations on the Aquarius Plateau and Straight Cliffs for surveying the topography.

G.K. GILBERT, POWELL SURVEY, BOX 806,
   WASHINGTON, D.C.
NOTEBOOK NO. 2
OPENED AT CAMP 12 NEAR SQUARE TOP BUTTE,
   UTAH, JULY 18, 1875
CLOSED AT SUPPLY CAMP AUGUST 8, 1875

{p. 1}   **Sunday, July 18, 1875**
Camp 12,   23.17 = 7730
Camp 12 is just on the edge of the G.M. [*Navajo Sandstone*]
SSE = ¾ᵉᵃ, 2 miles [*About superscripts see p. 14*]
WSW = 1½ᵉᵃ, 20 rods

These are measurements to determine the local dip and are made from a point on the Jurassic sandstone over the first gypsum. They indicate a dip of 1¾° to the north but the result is questionable on acct. of the short distance to the second point.

Thompson saw the Shinarump and the slaking shale in the DD Cañon

{p. 2}   so I must conclude that they are not absent in this vicinity as I had thought, but that I have failed to get down to them in my last section.

In the saddle of □ top butte is an outcrop of siliceous conglomerate, the pebbles of which are largely quartzite. It is nearly white on fresh fracture. 22.45 = 8800
Monday, July 19. Camp 13. Aneroid 4   22.08 = 9290
The chain of Aneroid 3 was broken last night.
Climbing the E salient of Aquarius [*Chokecherry Pt.*] the first outcrop is at 22.07 = 10,600 and is an ochrous shale with "*Chara*" and some tubular fossils. A little conglomerate over the shale.
Foot of the 'clinf' [*cliff?*] 20.55 = 11,200, An [*eroid*] 4. This is 100 or 200 ft above the base of the trachyte.
'From the crest of the salient' (ann. 4, 20.35 = 11,460) we see the whole face, but the great Howell fold would hardly be suspected.
The fold which runs SE from the Cedar Trachyte foothills is again shown in the duplication westward of the 'new' fold (W of the Twin Mesas). Perhaps the former fold turning a short angle 'is' the latter.

{p. 3}   The DD River valley running from the lava is at first monoclinal in the DD fold, then diaclinal across the DD Howell anticlinal [*Teasdale Anticline*], then cataclinal through the Howell fold [*Capitol Reef*]. It bears no consequent relation to the folds and so must be antecedent or superimposed. The presence of Ter. on Jura under M. L. and Aquarius and the intervention of Creta[*ceous*] at Table Cliff and Musinia shows a lack of conformity indicating that the Ter. is later than the Howell fold etc. and if this be so then the DD may be superimposed by a Tertiary system of drainage, i.e., by a system conforming to the slopes of the now eroded Tertiary. [*Gilbert noted that the Tertiary in Salina Canyon rested on Cretaceous formations. Finding the Tertiary on Jurassic*

formations at the Thousand Lake and Boulder Plateaus showed him that the anticline at the Red Gate is a pre-Tertiary fold.]

I need some names. These are provisional. Back of the "Belted Cliff" is first the "Little Cliff" and second the "Drab Cliff." The Drab Cliff stands at the W base of the Henry Mts. and surrounds the Twin Mesas. These are Needle Butte, Monogram 2, Monogram 3. [*Factory Butte and the North and South Caineville Mesas.*]

SE I can see the Howell fold for 30 miles and its section is Figure 2.1. The crest shows G. M. in the distance and near at

Figure 2.1. Looking along the Howell fold (Capitol Reef).

{p. 4}   hand and 'shinarump' and lower between the SW slope in the distance is 4° to 5° and the slope whatever its dip continues in actuality to Tantalus Creek at least. I now suspect that it continues as far as the fold is seen NW and that the SW dip of the round topped plateau is part of it, and that the complications of the region I have been studying are due to a cross system of displacements with ESE trend, together perhaps with a wave from the San Rafael arch which may be regarded as trending NE or NNE. The DD fold does appear beyond the Howell fold where it is manifested by inflection of the Little and Drab Cliff lines. The Little Cliff especially traces the modifications of the Howell fold due to the folds which cross it. It bends east toward the saddle of Henry Mts. Then after recovering its NW corner turns north along Monogram 3. NW at the crossing of the DD. N along Monogram 2 (where it is the New fold). W or NW to the San Rafael arch and there NNE along its flank.

A cliff along the gentle SW slope of the great anticlinal should show the details of cross displacement more definitely; but the trachyte by modifying

{p. 5}   erosion, prevents.
Still the fossiliferous Jura makes a contour at the base of the plateau and this holds for ten miles ahead. Camp 13 was only 2 hundred feet above it.
The Belted Cliff [*Morrison Formation*] opposite the saddle of the Henry(s) is plainly visible.
Camps 14. E. of E. Salient Aquarius.
    No. 4   21.74 = 9700
    No. 3   22.03 = 9300 (reset)
    No. 3 was disabled by the breaking of its chain and was

repaired by tying with thread, the chain to the long link by which it pulled the lever.

Comparisons before and after breaking.

Cist. No. 2144	A.T.	An. No. 3	Corrected Entered	Correction
22.184	51.5	22.520	22.184	-.336
21.995	55	22.020	21.989	-.031

No. 3 reads lower by 0.305 than it did before the accident.

Tuesday, July 20. Camp 14 at foot of E Salient of Aquarius. 21.84 = 9550.

Thompson says the general Aquarius Plateau is 100 ft higher than the E. Salient. The trachyte seems of the same depth as on M. L.

The Tantalus Creek bay seems to be walled about by palisade all the way.

Figure 2.2. Profile of the Escalante fold (Circle Cliffs uplift) as sketched by Gilbert from the East Salient of the Aquarius Plateau.

{p. 6}   Camp 15.   21.67 = 9760. Graves Station. 21.45 = 10040

It is now evident that the fold I have called provisionally Howell fold and "the great fold," should be entitled *The Escalante Fold.* It may be called a monoclinal limiting an inclined plateau or an anticlinal (Fig. 2.2). I guess its throw to be 4,000 ft. [*The fold, now known as the Circle Cliffs uplift—see p. 7 this notebook—has a topographic relief of about 4,000 ft; the structural displacement is about twice that.*] The Escalante Cañon cuts its SW slope and the Kaiparowits Plateau faces it. I read its SW slope today 3° in the 'xhan' distance (30 m) and it seems a little greater nearer at hand.

The Vermilion and Gray Cliffs are growing distinct but cannot yet be thoroughly discriminated.

30 miles away the Verm., there thin, completely covers the Esc. fold. It forms a cliff facing this way (NW). 5 m. from here it caps the fold in the same way and forms a companion cliff facing SE. These cliffs are connected by continuous (except for cañons) escarpments on the two slopes of the fold, the NE standing under the steep fold slope and the SW standing from 5 to 10 miles down the gentle fold slope. So there is a continuous cliff of erosion returning on itself and

{p. 7}   facing inward—the reverse of a mesa.

Inside the Vermilion the Shinarump circles in the same way. I will call the upper of these cliffs (the Vermilion) *The Circle Cliffs.*

The same thing geologically is presented around the Round Top plateau [*Miners Mountain*] and there it is due to an anticlinal. Perhaps the Circle Cliff has the same cause, the Esc. fold having a maximum or being crossed by an anticlinal in this region. It looks as though the G.M. plateau before me (bet. me and Henry Mts.) might be gently synclinal but it is to be remembered that another cañon exists for its preservation. It lies on the watershed between the Escalante and DD and is protected by the elevated Henry and Aquarius; so its base plane of erosion is elevated as compared to the Circle Cliff region. The preservation of the SE side of the Circle remains to be explained.

It is now evident, as Thompson before suspected, that the DD fold valleys find drainage across the anticlinal to the DD, another case of inconsequent drainage.

The foreground is G. M. except an occasional hint of Jura, and I stand on trachyte debris.

Point 119 dike 7° to the SE.

{p. 8}   The trachyte here and all about the slopes of Aquarius and on the Escalante is a dark gray paste with large xx of feldspar—it is the same as the sanidine 'dolinte' of Sierra Blanca.

Returning to camp and mounting Jan, I rode to a spur E of camp [*15*] where I had seen an exposure (Fig. 2.3). A few rods below me (E) is the G. M.

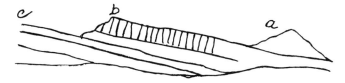

Figure 2.3. Section of spur east of camp 15 described on p. 8. The white sand (a) (Navajo Sandstone) shows above and the Shinarump (c) shows below the Vermilion (b).

Descending the hill the soil indicates red shale with cream, slaty calc sandstone and a bed of the latter is seen at 27.10 = 9220.

Although I am still above the G. M. I now project the SW wall of the Circle Cliff higher in the middle than at either end—a proof that the Escalante anticlinal rises higher there than at the ends of the Circle.

The hard bed is succeeded in 20 ft by another which contains a few specimens of *"Camptonectes,"* and massive x-bedded sand (a) begins at 22.17 = 9130. It is only 10 ft thick, has a shale of equal depth below it (b) (Fig. 2.4) then comes another x-bedded sand (c) and red shale and sandstone (d) to 22.25 = 9030.

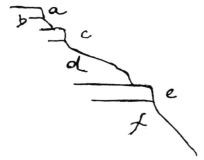

Figure 2.4. Section, probably in Carmel Formation on slope of Aquarius Plateau, described by Gilbert at bottom of p. 8 and top of p. 9.

Figure 2.5. Synclinal and anticlinal on surface of the G.M. (Navajo Sandstone) as seen by Gilbert from near the southeast corner of the Aquarius Plateau.

Figure 2.6. Profile of the Kaiparowits Plateau as sketched by Gilbert near his camp 16.

{p. 9}   The 30 ft of massive but not cross laminated, yellow sandstone (e) and the G. M. typical in (f) (9000).

In this attenuation of the top of the G. M. is a semblance of the Upper Paria Section. Looking N at the same horizon I read 1da. The distance is only 20 rods.

In the sandstone lettered C the cross bedding dips SE.

Camp 5 p.m. 21.63 = 9710

Wednesday, July 21st. Camp 15, on a small tributary of Pleasant Creek and near Mosquito Point.   21.67 = 9770.

After sketching the Circle Cliff found an exposure (above   21.23 = 10,310) of yellow sandstone with a band of chert and float of conglomerate. No fossils. This is ¼ m. W of Graves Sta 12. 1 m. N of Graves Sta 13 (21.17 = 10380) is a second outcrop of the same sandstone and 'clomerate'. There is more rock shown here and it dips NW 5° to 10°.

Any of these outcrops on the plateau slope may have slidden, but the occurrence of these two masses at the same height and a mile apart favors the idea that they are in place.

From a point near and above Graves Sta 13, I look southward over an ocean of G.M. Toward Navajo Mt. and the end of Kaiparowits it is all Vermilion limited at the left by the Circle Cliff. In the

{p. 10}   foreground are two or three miles of gray (beyond a mile of trachyte debris) and this forms a dip marked cliff parallel to the Circle. Its most remote visible part, however, is more red than white in color.

10 miles to the SSE the upper surface of the G.M. shows a synclinal (Fig. 2.5) and a companion anticlinal and the same undulation appears to cut the Kaiparowits cliff farther to the right. The trend may be SSW or SW. This side of the Table cliff there is a swell which runs out to the SE. It brings up the top of the G.M. as much as a thousand feet so that it hides the Kaip. Cliff. The latter dips down in its NW course as though the bed which makes it ran far below Table Cliff. These features are shown readily in the sketch (Fig. 2.7).

Camp 16.   21.60 = 9850.

After dinner I rode 3 miles SE to edge of the G. M. and took my seat at the fossiliferous limestone of the Jura (22.30 = 8970). I have the view of the morning from a lower station. The Kaipar-

owits has more shadow and shows plainly a shoulder, making its profile like Figure 2.6. The straightness of the plateau

{p. 11}   face shows that it is not a simple erosion line but is controlled by a displacement. I can see now that the wall is not unbroken. It is notched by several small cañons and by two large ones. The latter are not far apart and show S and a little W of S from here. I think the strata dip from the escarpment. The wall is higher at the SE than at the nearer end.

In line with the mouth of one cañon (S) and half as distant are two conical buttes standing on the summit of the G.M. Are they volcanic? (No.)

The appearance of the folds noticed this a.m. is not specially changed by the change of stations.

The topography of sandstone before me is controlled or rather influenced for many miles by the trachyte debris. All the flat tops are capped by that and slope from the plateau. It is a

{p. 12}   labyrinth of cañons.

The sandstone surfaces are many, often carved 'an tortoise'. Each block bounded by seams has a convex surface and the seams are cut in grooves. The cross lamination dips S and E.

Thursday, July 22.

From the hill near Camp 16 I have a sunrise view of Kaiparowits Plateau. It is built of dark and light bands and the light have the lustre of ivory. In the upper cliff all is dark at the SE except a faint streak at top and another near the base. The lower gradually increases westward and is joined by others until at the cañons the ivory is in the majority. At the same time the height of the wall increases. The lower wall on the other hand increases in height to the east and its ivory remains constant. The total of the two walls does not vary much. If these are the Drab and Little Cliffs I do not see the Belted nor can I detect any sign of displacement other than the straightness of the wall.

From Sta. 13 (Graves).

The butte I suspected yesterday of being volcanic is

{p. 13}   identical in color with the red upper G.M. beyond it and I now think that the green plain on which (it stands?) is not the top of the G.M. but is a plain smoothed by some erosion possibly that of Last Chance Creek. Perhaps but not probably the

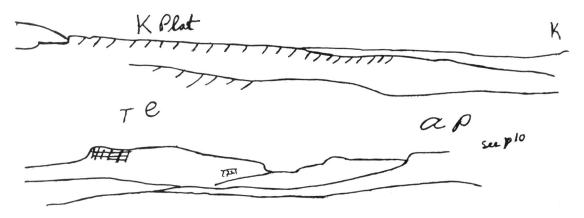

Figure 2.7. Sketch of the Kaiparowits Plateau, p. 16. (The Te in the lower section of the panorama probably refers to the Tertiary cap on the Table Cliff Plateau which forms the SW rim of the Escalante Basin. AP probably refers to the southeast slope of the Aquarius Plateau.)

monoclinal and anticlinal of that region are to be explained in the same way. We shall see.

Camp 16   21.60 = 9850
Sta 14 (Graves)   21.70 = 9730

Nothing new from this point. The upper limit of the G.M. is 2 m. S and is carried 2 m further south by the debris of the valley adjacent at the W, the valley of Camp 17—the outher salients, that is the reentrants are only a mile south.

The trachyte continues of the same coarse character.

Boulder Creek 3 m below camp 23.24 = 7850 (12 miles).

The first exposure is—

1)	Massive x-lam yellow sand,	15 ft.	23.13 = 7980
2)	Yellow and red shale	10 ft.	
3)	Massive 'aclon' sandst	25 ft.	
4)	Yellow shale	25 ft.	
5)	Yellow sectile limestone with *"Camptonectes"*	15 ft.	

23.08 = 8040   Dip W ½ m = 0. Dip S 1 m = 0.

6)	Yellow [*and red*] shale	15 ft.
7)	Yellow sectile (like 5 but more arenacious)	15 ft.
8)	Soft yellow green shale	50 ft.
9)	Gypsum, white	6 ft.

23.02 = 8120

{p. 14}

10)	Gypseous red shale	15 ft.
11)	Gypsum	6 ft.
12)	Red shale	25 ft.
13)	Hard gray shale	6 ft.
14)	Gypsum	11 ft.

22.96 = 8180

15)	Shale of many colors but chiefly green-yellow	100+ ft.

Friday, July 23, 1875. Creek at Camp 17.   22.69 = 8500

The gypsum (14) is exposed in Boulder Creek almost to camp and must lie there at about 8400.

In the escarpment west of camp the first exposure at 22.65 = 8560 is of 15+ that is seen continuously to the top about 400 ft. There is probably far less red than yellow but the red stain shows over all.

Camp 18.   21.52 = 9960

We have passed much jasper today and yesterday. It is of many colors and black. The black is all in rounded pebbles. The others are chiefly in large masses, angular. It may be a clay baked by the hot trachyte. The trachyte is slightly different from any we have passed and the constancy of the variety in any given hill, not plainly a terrace, is evidence of fact that the accumulation is 'merely' a settling down of trachyte under or nearly under the place of its first cooling. This limit of the trachyte then indicated is 2 to 5 miles outside the Plateau edge.

{p. 15}   **Saturday, July 24.** Camp 18.   21.42 = 10080
Sta. 14.   Graves.   20.93 = 10,710.

The fold which we are approaching rises rather suddenly on this side but falls as suddenly and further on the other. Its crest is not much more than 2 m. distant and is of G. M. Pine Creek cañon divides the anticlinal running partly along its creek and chiefly on this side. The top of the G.M. does not rise so high as sta. 14 by several hundred feet. Beyond the creek stands a ridge of Kaiparowits Cliff rock. Top Aquarius (x)   20.55 = 11,230.

Looking SW I see two folds crossing the Aquarius—two post-trachyte folds. Their trend is NW (Fig. 2.8).

Figure 2.8. Two posttrachyte folds cross the Aquarius [*p. 15*]. "No. 1 in Figure 2.8 is more distant by 5 miles than the Pine Creek anticlinal. No. 2 I can trace 10 miles."

No. 1 in Figure 2.8 is more distant by 5 miles than the Pine Creek anticlinal.

No. 2 I can trace 10 m.

We lunch on the divide bet. The Escalante and the creek that joins DD near Supply Camp.   20.68 = 11,030

{p. 16}   The folds now seem as in Figure 2.9. Where we just climbed we stood on *0* and 5 rose 500 ft above. Finally near the divide *0* is lost being merged in 5. 5 is the fold which runs W of M. L. (or 5 and 0). Its general course is W of N but it is convex to the W, slightly. Where we just struck it the trend is nearly NNW.

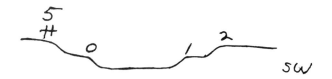

Figure 2.9. The folds as seen from the divide between the Escalante and the creek that joins the Dirty Devil near Supply Camp.

Climbed a spruce. 'Fold' No. 2 diverges widely from 5, trending WNW or NW and ends visibly at Graves Valley [*?*]. No. 1 cannot be traced so far. The whole country between 5 and 2 undulates with a NW trend to the rolls.

Returning—I note 10,950 at base of No. 0 and 11,300 its crest at its base a large sink hole (or rather two) like the Virgin well.

Camp 19.   22.06 = 9280

Camp 19 is in G.M. inclined 15° to 20° and the ridge separating it from dry valley at the E is a monoclinal capped by fossiliferous Jura or the sandstone above it.

Sunday, July 25. Camp 19   22.11 = 9210

at 21.38 = 10,120 the top of the sandstone and of the fold.

{p. 17}   Point 15 (Graves)   20.77 = 10,920

Fold No. 19. A mono with E dip runs SSE 10 m and then is lost to sight. It brings the sandstone outcrop to the foot of this cliff but leaves some islands of Jura. It runs through camp 19. The companion mono. No. 20, may have the same trend or one more nearly SE according as it cuts the Straight Cliff near Potatoe Cr or beyond False Creek. Perhaps both these monos unite in 20. No. 20 passes 2 m. SW of this point.

From the cañon of False Creek a synclinal runs at first S and then curving to the SW, passing entirely E of Kaip. peak. It holds nearly to the Straight Cliff—superior rocks. To the right of it is an adjacent anticlinal and then another syn. The whole seems to be combined with an anti. running E or SE a little way back from the Straight Cliff and between that and KP.

I see no sympathy between these folds and those of the plateau. No. 1 is indefinite and may not be a fold. There is a step of 100 or 200 ft from the lowest level of yesterday's diagram down to the level of Sta. 15 but its 'counlone' is irregular and suggests a 2d eruption after some erosion. The lava here is coarse grained like that of the Flagstaff and it is a thin sheet not over 200 ft

{p. 18}   and probably much less. The creek south of this point runs in a great cañon that looks as though it might be cut out of this table.

No. 2 is still definite and trends nearly W. Its E end is just beyond No. 20 and its throw is 400–500 ft.

The creek cañon S shows Jura all along its further bank and at one point on the nearer.

A minor step in the Sta. 15 bench—a step of 50 ft—is due to the failure of the upper of the flows composing it to cover the lower. This helps to the belief that the bench is due to contemporaneous erosion. On the other hand a strong evidence of displacement is shown at the point Prof. Thompson and I climbed (Fig. 2.10).

On the next page is a sketch to show the naves of trachyte due to sliding.

Note. The bearings from this point are all based on the assumption that K.P. [*Kaiparowits Peak?*] is due S local attraction.

Figure 2.10. Probable displacement at point climbed by Gilbert and Prof. Thompson (p. 18). The high plateau (a) dips off both S and W as though one displacement (d-e) had crossed another (b-c). f and g are on a level with each other. The line c-b continues W toward b for several miles and it looks straight but it is lost at the canyon S of Sta. 15.

Figure 2.11. "Looking NE from 15 Sta. at the spur on which was camp." Henry Mountains on skyline.

{p. 20}   Noon camp (22.40 = 8860) is on trachyte debris, but beneath is Jura of the 20 fold. We left the G. M. when we crossed the creek 3 miles back.

From the point at left of the trail near the divide ¾ mile from camp I see the Straight Cliff series inclined by fold No. 20 to the SW. It must be Cretaceous or No. 4 of Howell (Fig. 2.11). The valley below there is mono. by accident and will give a good section.

The fold No. 20 must turn or subside, for looking straight down its monoclinal valley I see it for 5 miles and then beyond the gap and 5 miles further stands the Straight Cliff unaffected

Figure 2.12. Cross section of the monoclinal near camp 20 (p. 21). The fossiliferous Jurassic and associated beds are eroded and covered by the floodplain at camp but appear in a hogback a few rods NW. At *a* is the heavy gypsum.

{p. 21}   by it. I think likely it turns to the right (S).
Camp 20. 23.48 = 7570
Monday, July 26, 1875.
     [*Fig. 2.13 is a cross section apparently near camp 20.*]
T1 = yellow and gray shale   30 ft
T2   yellow massive sandstone with 'shiny' stones   20 ft
T3 '   ' shale   10 ft
          T from top of T2 = 33½^{ea}
T2 dips 16°
T4 = massive coarse yellow sand   20 ft
T5 = gray shale with sand and lignite   20 ft
T6 = massive yellow sand   20 ft
T7 = gray shale with 6 in coal   6 ft
T8 = massive yellow sand   15 ft
T9 = gray and yellow shale with silicified wood (and some sand)
          50 ft
T10 = massive yellow and white sand   40 ft
T11 = yellow shale   10 ft
T   = massive yellow sand   10 ft = 22.85 = 8300 = 9 a.m.
          T dips 8° and strikes NNW. At the base of the T series saw *"Inoceramus"* in loose stone. It must have come from the escarpment.
     From T to U the rocks are covered. At U is yellow sand 10 ft exposed.
U = 22.95 = 8200

Figure 2.13. Cross section, apparently near camp 20. T and X from S = 15 ½^{ea}. S = 23.18 = 7920 = 8 a.m.

{p. 22} U^1 = a deep arroyo showing yellow shale = 22.'76' = 8180 (Fig 2.14).
Section at Figure 2.14:
V^1 = soft shaley yellow sand   30 ft
V^2 = massive yellow and red sandstone. *"Ostrea."*   75 ft
Top V^2 from base V^1 = 36^{ea}. Dip top of V^1 = 8°.
Top of V^2 = 22.70 = 8500. Dip of V^2 ½ m. NW = 6°)
) 8°
Top of V^2 = 22.70 = 8500. Dip of V^2 20 rods SE = 9°)

Figure 2.14. Cross section at U in Fig. 2.13. "V from U^1 = 32^{ea}. Dip at V = 11°. At U^2 are 5 ft. of yellow calcareous sandstone intercalated in yellow shale. 22.875 = 8270 = 9.40 a.m. V from U^2 = 32^{ea}. U^3 = yellow shale. 22.82 = 8340."

The dip increases to the SE. Strike of V^2 = NW and S.
V from V^2 = 34°

V^3	gray shale	5 ft
V^4	yellow massive sand	25 ft
V^5	gray shale	5 ft
V^6	soft yell. mass. sand	10 ft
V^7	yellow and gray shale with "Ostrea"	30 ft
V^8	soft yellow sand	25 ft
V^9	bit. shale	4 ft
V^{10}	yell. sand and shale	20 ft
V^{11}	massive yellow Ss	25 ft = 22.51 = 8720.
	V^2 fm V^{11} = 30da	
V^{12}	yellow gray shale	45 ft
V^{13}	mass. yell. sand	15 ft = 22.35 = 8920
V^{14}	shale	20 ft

V^{15} = V = mass. yell. sand. Locally red and purple at top (shark's tooth)   22.26 = 9020 = 11 am. 60 ft

{p. 23}   The G.M. to the NE is a few feet (50) higher than V. or 1550 above camp. Its highest point is its nearest and it loses very slowly to the N more rapidly to the SE. Walter is right about Agate Creek. It enters the G.M. where we did not dine and comes out at Camp 20 passing so close to the fold that one wall has dipping beds all the way and the other the level beds most of the way.
Dip of V = 5°

X	=	17½^{ea} from V		
X^1	=	shale	50 ft	
X^2	=	mass. yell. sand	50 ft	22.18 = 9110 11.20 am.
X^3	=	yell. and gray shale	40 ft	
X	=	mass. and bedded sand with sand concretions 100 ft		
X	=	22.08 = 9230		
Y	=	yell. and gray shale and yell. sand in alternation (Fragment of bone)		

Hilltop = 21.95 = 9410 = 11.45

To the SW the beds all dip SW for some miles. 2 m. away beds which are probably superior 'trituse' dip 4°. They are just beyond the first creek. It is not over two miles W to the trachyte debris but there are exposures beyond the beds several hundred feet higher than this. I can see the fold parallel to this at the base of Table Cliff. Toward Last Chance Creek V has a very slight dip not over 2°, to the SW and X is nearly level.
(a) Nearer by, just beyond False Cr Cañon both cliffs (bb)

{p. 25}   dip more strongly to the SE. The elongation of the lower cliff at that cañon indicates that a synclinal leaves the plateau there but there is no other evidence of it from this point. From False to Plateau Cañon the upper cliff describes an anticlinal and synclinal (cd) and the mono between them runs here (ef). The trend of the synclinal is S by E from here to Potatoe Cr; then curves to SE or even ESE and then curves quickly to S. On the line of Potatoe Cr it is succeeded by an anti (g) and this I can detect at no other point to the right but to the left it probably makes the swell (h). This would make it parallel to the synclinal.

At K.P. I make out nothing and of Table Cliff little except that it holds Pink Cliff with some trachyte above.

Near foot of Potatoe Valley there seems to be a synclinal dividing No. 20 but it may be only trachyte debris.

Returning to X at 1 pm 22.10 = 9220.

The series Y (p. 23) has many ferruginous hollow and lining or coating concretions and fossil wood, partly in iron and partly in silex.

{p. 26}   Angles from X (Fig. 2.16)

K =	2da
M=	'4' ½^{da}
O =	6¼^{da}
P =	7½^{da}
R =	10da
T =	13¼^{da}
V =	16-2/3da

It is safe to call the whole of V yellow shale. In it I find fragments of the large "Inoceramus" and a fragile small oyster adhering to it.

Returning to S at 2.50 pm = 23.18 = 7920
R = 23.12 = 7970
P = 3.10 pm = 8170

From R to P the only exposures are gray and yellow shale with "Gryphaea."   O = 22.90 = 8250

From P to O there are no exposures except that the shale

{p. 27}   continues a few feet and there is a hint of yellow sand. O is covered by trachyte and white gravel.

The white 'grove' proves to be a conglomerate = Oa = 20 ft and Ob white shale 10 ft, Oc = massive sandstone '   ' friable that it makes steps of sand—23.08 = 8040 ('8150, p. 32']
Hail
Od   Beyond this '   ' it '   ' formation is '   ' an arenaceous shale, red with some white.
N = 23.12 = 7880 = 4.15 pm. = dip 31°.

From N to N^1 the color is red only.
N^1 = gypsum = 23.27 = 7810
M from N^1 = 14^{ea}

Figure 2.15. Unlabeled view sketched on p. 24. View is SE and S from the south side of Boulder Mountain. Escalante River Valley at right?

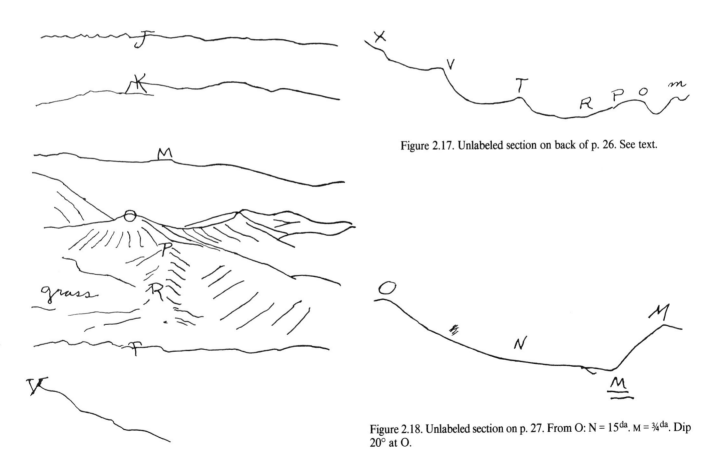

Figure 2.17. Unlabeled section on back of p. 26. See text.

Figure 2.18. Unlabeled section on p. 27. From O: N = 15da. M = ¾^{da}. Dip 20° at O.

Figure 2.16. Unlabeled section referred to on p. 26. See text.

There is a second band of gypsum and interc. red and white shale to M

M = 4.35 = 23.20 = 7790. M from M = 24_{ea}; dip 28o.

M = 22.99 = 8150 = 5 pm

The summit of M is the foss. Jura (dip 25°) or

{p. 28} the sand above, probably the latter (Fig. 2.19).
The ' ' G.M. is as above. Total 200 ft?
Pine
K from M = 15^{ea}. J (beyond Oyster Cr) is invisible.

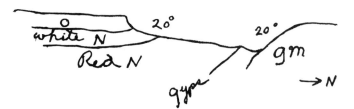

Figure 2.19. Sketch of fold with G.M. (Navajo Sandstone), p. 28.

The plant found in *T2* was seen also in a number of superior beds, up to V. Its range thus includes the *"Inoceramus"* of V3 and it helps to identify Howell's No. 4 in Salina Canyon where also it is found.

Where the fold begins the G.M. seems bent ' ' suddenly. The dip is ' ' halfway from Camp 20 to Camp 21, and its max is 35°.

An old level of the flood plain is finely marked by a clean planing of sandstone shown for a long distance in the creek bank. Other and higher levels are marked by red bands on the sandstone—another illustration of the accidental nature of the red color of parts of the G.M.

{p. 29} **Tuesday, July 27.** Camp 21 = 24.40 = 6530
Pine Creek = Oyster Creek cuts into the G.M. before joining Birch Creek and the two united 'rund' straight into the fold. [*Stream names in Potato Valley area are confused. Gilbert's Pine Creek probably is modern Pine Creek but Birch Creek remains*

unidentified. Gilbert's map (Fig. 2.A) shows these as Winslow Creek, a name not used in his notes. Boulder Creek on his map probably is the modern one of that name.] The last part of the cutting is clearly superimposed by the trachyte drift slopes and the whole may have been. Trachyte boulders are scattered over all rocks of the region. I found them yesterday on my summit 1500 or 1600 ft above present drainage lines. The point on which I sit (24.14 = 6810]) is made of waterworn assorted trachyte and a bench of the same lies a quarter of a mile beyond in the cañon. I see two remnants of a bench several hundred feet higher. The story of False Creek is to the same pinpoint.

The G.M. dips here 20° and a mile to the SE the Jura lime and gyps creep up 400 ft on its flank and limit the view. The top of the G.M. is about 1200 ft above water and dips S and E.

Near Camp 20 the O sandstone is harder than described yesterday and might be taken for the G.M. if nothing more was seen. The hardness is only local. Farther downstream it resumes its character and 'gives'.

It is now plain that fold

Figure 2.20. Section at Pine Creek. The summit of G.M. (Navajo Sandstone) is 1300 ft above Pine Creek. The summit of O is 600; the base of O is about 250.

{p. 30} No. 20 [fold no. 20, p. 17] does divide. I [*it?*] passes under the table of trachyte drift opposite Camp 20 as one and emerges as two. The Pine Creek fold runs to this point straight and turns a mile SE. At a point a mile NW its section is as Figure 2.20.

The spur of foothills between Birch and False Creeks is not Lower Straight but is O, but the two so simulate in color that I cannot yet define their outcrops. The level O of the section in Figure 2.20 holds SW for two miles forming a SE facing cliff and then begins to dip gently to the SW in the second bench of No. 20.

The T beds at False Cr. are represented by a single yellow sandstone (24.10 = [?]) capping shales gray and black and green. Perhaps the black is coal. Dip 8° to the SSW (Fig. 2.21A)

{p. 31} 'Sta 17'
Hilltop 23.20 = 7900 = 12 N.
The top of the hill is 2/3 as high above U as I climbed yesterday. On the way up and at the top I found *"Ostraea"*

Figure 2.21A. "The T beds at False Creek are represented by a single yellow sandstone capping shales gray and black and green. Perhaps the black is coal. Dip 8° to the SSE." (Notebook 2, p. 30) Presumably this is the Dakota Sandstone at base of the Mancos Shale and top of the Morrison Formation.

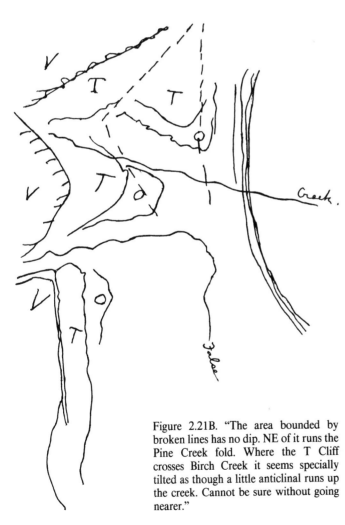

Figure 2.21B. "The area bounded by broken lines has no dip. NE of it runs the Pine Creek fold. Where the T Cliff crosses Birch Creek it seems specially tilted as though a little anticlinal runs up the creek. Cannot be sure without going nearer."

(large), a large ribbed concifer *"Bacculites velontinia"* (?) leaves etc.

The shales U are all gray and 'weather' yellow.

The bed I am now on dips 6° to the S and strikes W (1 m) and ESE (40 rods).

The synclinal and anti. S from here look just as yesterday. 'This' rocks superior to the Straight Cliff are yellow and sandy (?) and there are 1000 ft of them in the synclinal and the Castle Buttes.

The T cliff has a promontory between False and Birch Creeks but goes only half as far as the O cliff.

{p. 32}   The beds marked *White N* in the sketch on p. 30 are white massive sandstone and 'glimmer' under the O cliff 3 miles farther up Birch Creek and SE along the base of the lower cliff.

The fold which I have been watching from a distance for a week seems to be a prolongation of the 19–20 *en echelon* as drawn in Figure 2.21B. The dip from *a* to *b* (from this fold to Straight Cliff) seems continuous and gentle—almost zero at the Cliff and barely perceptible in the lower cliff.

From this distance the highest point of the Escalante fold seems between the large Shinarump Island and the SE Circle Cliff.

The Tertiary of Table Cliff extends under Aquarius until cut off by the 20 fold.

Descending found Howell's *Inoceramus"* at top of lowest V sandstone. No good specimens, valves all singles.

{p. 33}   Section between False and Birch Cr. Beginning with T.

The shale V is so greatly developed that I suspect it has swallowed some of the T.

The beds are dipping 7° to 8° and the section has to be made in driblets.

3:30 pm = 23.93 = 7050		
T = yellow sandstone		15 ft
2 = yellow shale, *"Ostrea"*		.5 ft
3 = Bit. shale		1 ft
23.95 = 7030		
4 = gray aren. shale		8 ft
5 = coal		2½ ft
6 = Bit. shale		2½ ft
7 = gray and bit. shale		15 ft
23.975 = 7000		
7½ = white shale and sanstone abruptly ending		10 ft
8 = green shale		20 ft
9 = red, purple shale		20 ft

The purple shale has 'concretions' of limestone and slates.

24.'02' = 6950

10 = yellow and green shale		10 ft

A little beyond 7½ reappears and 9 is replaced by a mass of chert.

Then under 10 comes

11 = a coarse ochrous sandstone. The change from shale and lime concretions to chert is so abrupt that I suspect the chert is a slag from a burning coal bed.

Upon the sandstone 11 are some buttes capped by masses of conglomerate not elsewhere seen. It equals T or 7½.

{p. 34}   beginning at a new spot 23.93 = 7050
No. 11 here a fine conglom   20 ft
12 = red shale   20 ft
13 = white sandstone, bedded coarse
       24.03 = 6940
14 = red shale   30 ft

Wednesday, July 28, 1875. Camp 21 = 24.29 = 6650
Section in right bank of Birch Creek. Ascending.
Dip 22° to the SW

1 = G.M. cream cross laminated sandstone   1,200 ft+
2 = bedded yellow sandstone   15 ft
3 = red shale and sandstone   30 ft
4 = shotted with ferruginous concretions   15 ft
5 = massive xlam pale red   20 ft
6 = gypsum   15 ft
7 = pale gray shale   20 ft
8 = cream sectile sandstone dendritic and
      calcareous, rippled; partly a limestone   30 ft
9 = red shale and gypsum (¼ gypsum)   40 ft

Near the crossing of the Last Chance Creek the

{p. 35}   whole Jura is exposed (Fig. 2.22). The cream dips 5° and everything else 2° at most.

A little farther where we climbed out of False Creek, valley No. 3 exhibits a dip of 10°.

Near Last Chance we travel in No. 3 and encounter a group of eroded buttes the necks of which are dark brown while the caps and bases are paler. Cross bedding gives an oblique slope to the caps and contributes to the monoclinal appearance.

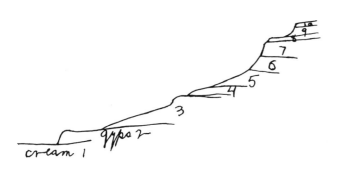

Figure 2.22. Section near the crossing of Last Chance Creek. "3 = red soft sandstone = 'elsewhere' shale; 4 = pale gray shale; 5 = yellow soft sand; 6 = white mass. and lam. sand; 7 = red and gray shale; 8 = thin yell. sand; 10 = thin yell. sand; 'retired' ¼ mile on the foot spur."

The foot spur next to L. C. Cañon owes its exceptional size to the thickening of No. 8.

At Camp, No. 10 is massive yellow and cross bedded not distinguishable from G.M. inclination. Dip 4½° to the SSW. There is the same dip across the creek. I measured a single bed 55 ft thick, the limits of the bed being shown by the

{p. 36}   cross-lam. The full exposure is 100 ft and that is not the whole of No. 8.

Saw the creek come down. The bore bore pine cones and leaves and sticks and was a foot high and the water rose very little after it passed.

In clambering over the smooth sandstone we found 34° the limit of the slope we could ascend with our hob-nail shoes.

Cross bedding dips SE. Greatest dip noted 26°.

Thursday, July 29. Camp 22 = 24.67 = 6210.

The shale under T is 40 ft thick and has '   ' beds and oysters. T is 30 ft thick—a soft yellow and shaley below.

The gray shale is several hundred feet thick and fossiliferous.

There is no trachyte in the drift of L.C. Creek nor did I observe it on the Sta (17) near False Cr.

At 24.66 = 6340 found *"Mammites"* in the gray shale about 100 ft below its top.

Ascending the stream until 10.20 am (24.35 = 6570) we passed 2 or 3 of the heavy sandstones and then found a good coal horizon. The shale interval between sandstones is about 75 ft and

{p. 37}   in this are three beds measuring 5 (above), 12, and 7 + ft separated by gray and bituminous shale. At lower horizons are other bit. shales. Just above the coal I saw in the sandstones fragments of the large prismatic *"Inoceramus."* Above the workable coal are two thin seams—and a mile farther on we find two others. Still another blossom shows a hundred feet above dinner camp—and another—a third at 23.72 = 7070, and with it a zone of profuse iron concretions like those of Alphabet Peak in fold No. 20.

Hilltop 23.72 - 7300.

There is a slight general dip to the WNW—parallel to the cliff which we have left 5 miles behind. Besides this there are a gentle synclinal and anticlinal revealed by the cañon wall as in the diagram (Fig. 2.23). The dip at mouth of cañon is not over 6° or 7°. Here is it zero or 1° or 2° to the ENE and ahead it may reach 10° but is more likely 8°. The cliff 4 miles ahead is of rock I have not seen. It looks much like this and lies on the synclinal which starts on Birch Cr., 2/3 of this section is sandstone

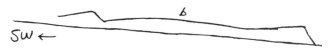

Figure 2.23. "Gentle folds revealed in the cañon walls."

{p. 38}   and the rest shale and coal.
    Dinner camp = 24.20 = 6750.

There is a distinction between the rocks of the Cret. and those of the Jura and Trias that is easily perceptible if it cannot be accurately defined. In color the Cret. sands have a yellow verging on green; the Trias lean toward orange. The Cret. usually are bedded but this is not diagnostic for they are sometimes massive and x-lam. The Triassic rocks abound in bright colors and strong contrasts. Their shales are often red and usually gypsiferous. The Cret. shales are gray and yellow green and they bear coal. The Jura and Trias are closely allied. Their sandstones and to a great extent their shales agree in characteristics. They make 1 great group well distinguished from the Carb. and Cret. The Cret. rocks resemble the Ohio Coal Measures in their color and texture of their sands and shales but lack the streaks of limestone.

In a mile we meet the base (after a shower) and have passed five veins of coal ranging from 1½ to 3 ft in thickness. We see two others within 75 feet in the cliff and the lower must be 6 or 7 feet thick. (No it is black slate with 2 ft of coal.)

{p. 39}   In leaving D in Camp we seemed to pass an anti-[cline] but the seeming may have been due to a western bend.
    A red shale suggests a burned coal bed.

'Horse' = John Thomas = 24.00 = 6960 = 3.20 pm.
First bench = 3.35 pm = 23.85 = 7150
[*A topographic profile, apparently looking S and SW here is crossed out. I do not recognize the two buttes indicated.*]
    Top of cliff = 4 pm = 7620. Top cliff = 4.20 = 7990 = 23.12.
    (See p. 41) Looking forward I see the synclinal parallel to its face running into the end and becoming deeper. At its very end it shows only its W dip and the point beyond is level. There is no great dip here. This cliff is rock entirely similar to that below. Last Chance heads against it. The other branch is larger and diverges.

Figure 2.24. Cross lamination in Cretaceous sandstone, referred to on p. 41.

To the SW there is a dip of 2° for 2 or 3 miles and then the strata either become level or rise (large cañons rise at the SW end of the butte and cover more space for 4 miles than the remaining rock). I think they rise a little and that the synclinal is continuous southeastward with one that is conspicuous 10 m to the S. There is likewise a change about here in the 'ground' dip. South it is to the north as in the Kanab region and continuously with that region; but to the west it seems to be southward.

The synclinal bet. this point and the Straight Cliffs is not the mono back of sta. 17 but echelons with it, just as the antis. do in the valley below. The gentle anti. between the two is probably the one at camp. The Castle Butte synclinal is the more decided of the two.

Near camp I find cross-lamination with a dip of 10° in gray and black arenaceous shale with fillets of coal [Fig. 2.24].
    Camp 23   24.08 = 6880
    Figure 2.25. July 29, see p. 39.

{p. 42}   **Friday, July 30.** Camp 23 = 24.30 = 6640
    Climbing a more easterly point of the spur I was on yesterday I find bivalves near its base. I am on the SW edge of the anti. at *a*
(p. 41). The crest is at that point in the sketch (Fig. 2.25) and runs to camp, the trend being SE. An equally gently synclinal lies SW of it; the same syn. that I saw yesterday,—but it has passed its

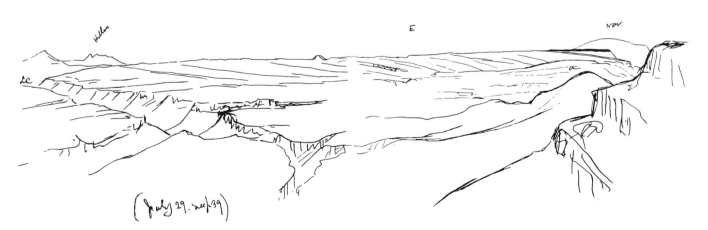

Figure 2.25. View is east across the west flank of the Circle Cliffs uplift from top of the Straight Cliffs. Nav = Navajo Mountain.

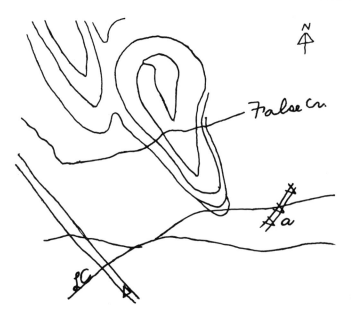

Figure 2.26. Sketch map of folds crossed by False Creek.

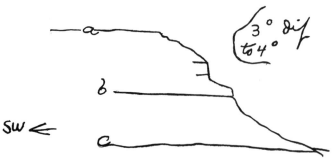

Figure 2.27. "This is profile of the Straight Cliff at Last Chance, *ab* = *cd* in depth = 1,000 ft or 900. *ab* is chiefly sandstone; *cd* is the gray shale." [*Should be* bc*?*]

maximum. I think it runs out in 8 or 10 miles and its place is taken by one which divides the anti. at its left. It is a sea of gentle billows.

Dinner camp 23.89 = 7090. We are on a cañon leading to 'Traverse' Creek and are a trifle above the horizon of the morning's camp. The dip is not great whatever it is and I suppose we are at the base of the syn. nearest the Straight cliff. Here is a leaf of the same species we found at Sta. 17 and this morning I found *"Sabal"* at the top of the cliff near Camp 23.

On a peak of the K. Plateau near camps 22 and 24 (22.98 = 8150 = 6.30 pm).

This is the usual capping bed of the plateau.

I can see that the rocks between me and the Escalante fold are in waves, but I cannot define them. The one that False

{p. 43}   Creek cuts is bent down suddenly where that creek crosses it—bent eastward (Fig. 2.26).

There is a mono. at a trending toward Mt. Hillers but I cannot follow the beds which reveal it nor make out its relations. Red is so prevalent and the exceptional white or yellow so fickle that in this twilight I can see little but the maze of cañons,— beyond L. Chance and toward the east.

All the folds below are gentle swells compared with that great base, the Escalante.

Coal blossom at the top as everywhere else.

From Dinner Camp to the plateau edge the strata rise and we rose with them. The latter point is on higher rock than the former.

The drainage of the

{p. 44}   plateau, from L.C. Creek eastward seems to be consequent. The main cañon in which we dined—a tributary of Warm Creek—follows a synclinal as far as I could trace it and its side gulches on the NE run regularly and directly from the cliff.

The Cliff is steep. I measured its inclination from summit to edge of lowest wall—47°. It is a divide and the cañons leading SW have been cut away by the receding cliff so as to be exhibited in cross section.

Saturday, July 31. Camp 24 = 23.46 = 7600
Plateau top near camp 24 = 23.28 = 7800.
Sta. 18 Graves = 23.03 = 9 am

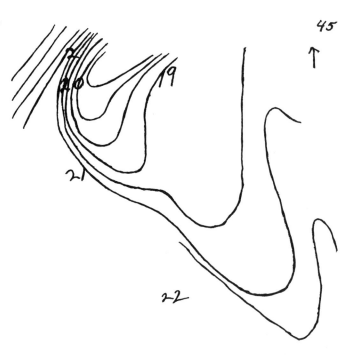

Figure 2.28. This drawing on p. 45 is without a caption. It is a large-scale map of the west part of the map shown in Fig. 2.29.

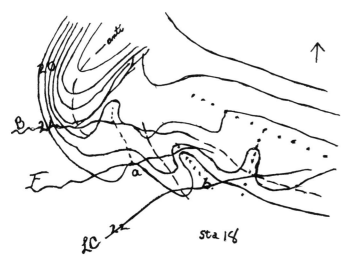

Figure 2.29. "Map of the anticlinals and synclinals between Escalante fold and K. Plat. = between Circle Cliff and K. Plat. The crest of Escal. fold from the Circle Cliff to the Colorado diminishes in height. The syn. at *a* carries Jurassic to the second contour. Syn. *b* carries a narrow cap to the 2d contour" [*compare Fig. 2.28*].

Figure 2.30. "K. [*Kaiparowits*] Peak shows a pink face (p) marking its crest as Tertiary."

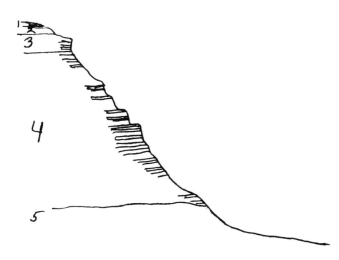

Figure 2.31. Section at Sta. 18, descending, Sta. 18 = 23.00 = 8130 = 10.45 am. (1) Sandstone, yellow and lam, 10 ft; (2) soft unseen, 30 ft 23.05 = 8080; (3) massive sandstone, 80 ft; (4) sandstone and shale; (5) gray shale 1 to 4 incl. = 1200 feet by Aner. 6".

{p. 46}   **Sunday, August 1.** Camp 24 = 23.35 = 7720

At 23.53 = 7520 the massive sandstone carries the large *"Inoceramus"* and an *"Ostrea."*

At 23.86 = 7120 is a streak of conglomerate with *"Ostrea."* The top of the section at this point is about 100 feet higher than camp. The wash of water from the camp arroyo has barred the sandstone ledge so that I can see them very plainly.

At 7120 a 'scuti' root in place.

Camp 25 = 24.68 = 6220

Monday, August 2. Camp 25 = 24.66 = 6230 = 7.20

{p. 49}   I suspect the mono. I saw between bet [*?*] Sta 18 and Hillers is not a true one but only a 'seeming'.

Last night I found coal in *c*—a thin seam close down to *d.*

Looking back from our change station I see no fold between the Escalante and K. Plat. There is one dip all across, a dip of 2°–3°. The Vagabond sandstone has become soft as shale and so has the Mushroom sandstone. The conglomerates of Camp 25 have increased [*or is it decreased?*] in their thickness, one dingy with the stain from the shales between the rest on a chocolate and gray shale. The Straight Cliff is accented now in the middle and a little way ahead the summit is rounded (Fig. 2.34).

Lunch at 25.02 = 5860 = 300 ft below Camp 25 horizon.

If there is an anti. bet. here and the Escalante it is under the plain the soft beds.

At 25.00 = 5850 we begin to climb. At first red sand

{p. 50}   as soft as shale.

24.80 = 6070 = 3 pm

b) Chocolate or purple and white sandstone—a continuation of the above but hard enough to make a cliff—bedded. At the very top is c) a thin bed of yellow sandstone (a part of ' ' is conglomerate).

24.12 = 6840

There is a dip here that I guess to be 5°.

Everything is red from *a* to the Escalante. Beyond that river the Gray Cliff sandstone is gray.

The slope we have passed today is a good place for "Witch" pinnacles (Fig. 2.35).

Opposite the point we climb is a red pinnacle 5 m. out in the valley, slender and inclined and probably as large a scale as the tower of Pisa.

First rocks of the upper slope at 23.63 = 7400.

{p. 51}   **Tuesday, August 3.** Camp 23.98 = 7070.

Hilltop—Ledge of K. Plat = 23.14 = 7960 = 8 am. The highest ledges of all are 40 rods back and are probably 40 ft higher. There is little or no dip. If any, the drainage indicates that it is southward, or from the edge.

End of K. Plat = 23.00 = 8.30. The section is in Figure 2.36, A and B.

No. 3 extends 3 mi. SE in a spur—unclimbable—and flanks the SE spur of No. 1. Between them is No. 4 and that covers the plain to the NE, E, and beyond the river to the SE. The Escalante

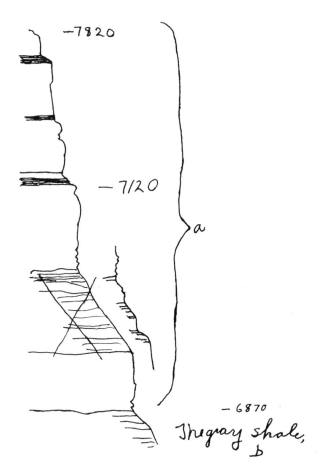

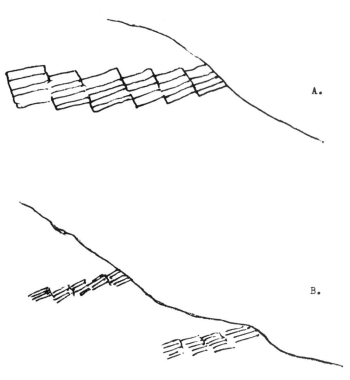

Figure 2.33. "The manner in which an intercalated sandstone has settled is notable. It everywhere dips into the hill at an exaggerated angle (A) and has slidden in detachments as sketched in B."

Figure 2.32. Section at Straight Cliffs east of Last Chance Creek. "a = yellow sandstone, soft, massive, heavy thedded with partings of gray arenaceous shale and shaly sandstone, 24.09 = 6870. b = gray shale, weathering yellow-green and perhaps of that color at top. Yes it is yellow at top and contains sand. The fine gray begins at 24.20 = 6750.
Allowing for dip the base of b is 24.78 = 6100.
Looking across from c to

{48}   foot-spur beyond L. Chance I make its dip evenly 5° (instead of 8° as before).
        24.40 = 6520
c = yellow soft sandstone and yellow aren. shale with oysters. The shale locally predominating. This is the coal horizon of Brush creek. On lithological grounds I should call it the base of the Cret.
        24.45 = 6460
d 1) conglomerate or coarse sandstone, white and ochre yellow. This caps the foot spurs usually. With slight interruption by red and yellow shale, it extends to 24.50 = 6570.
d 2) gray shale - 5 ft
d 3) red shale - 10 ft
d 4) white sandstone - 15 ft
d 5) red shale - 20 ft
d 6) white sandstone - 5 ft
d 7) red and yellow aren. shale - 30 ft
        24.65 = 62.40
E = white to yellow massive xlam. sand (vagabond Haven Rock) - 100 ft
Vag. Haven = 1 pm - 24.70 = 6190
The preceding section from c to e (incl) was made on the strike and needs no correction for dip."

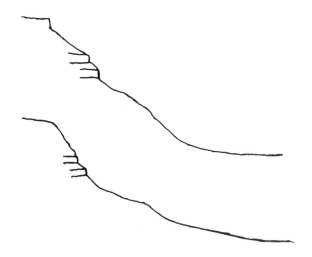

Figure 2.34. Unlabeled drawing apparently to illustrate rounding of the summit of Straight Cliff referred to on p. 49.

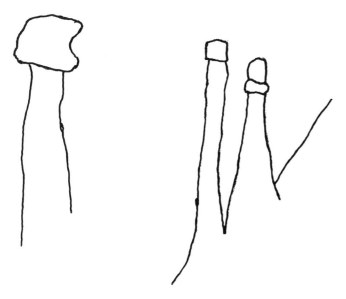

Figure 2.35. Witch pinnacles. "The diagrams are not exaggerated. The pedestals are gravels of the slope."

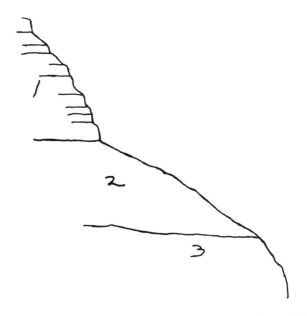

Figure 2.36A. Section at end of Kaiparowits Plateau. 1 = Cret sand; 2 = Gray shale; 3 = Chocolate or purple and white conglom.

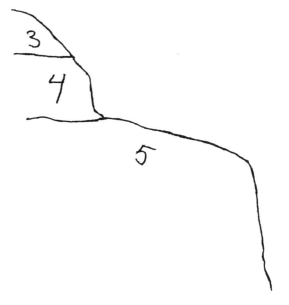

Figure 2.36B. Section at end of Kaiparowits Plateau, continued. 4 = red soft sand, passing into; 5 = pale amber sand = the Gray Cliff and Vermilion—extending to the river.

fold partly runs into Navajo Mt and may partly swing to the left and join a syn. that trends ESE beyond the river.

Navajo Mt. (a shade E of SE) on all its visible flanks, from ENE around by N and W to the SW is built of these rocks dipping away from its center. It is truly a volcanic cone of elevation. [*A structure contour map of Navajo Mountain is given in Figure O.E. The stock that produced the dome is not exposed. The dome is same area but lower than the one at Mt. Holmes.*] I measure 11° at the NE and 15° at the SW on its flanks— dips that carry the Trias almost to its crest; but its crest is dark with lava. [*The capping rock on Navajo Mountain is not lava; it is sandstone belonging to the uppermost Jurassic and lowermost Cretaceous.*] No. 4 on the Navajo is white. My view to the south is limited by a spur of the K. Plat. but to the right of Navajo I see some islands or formations of No. 3 No. 4 has an island on the SW base of Navajo, an island with dipping strata. It has also a small island on this flank. The chief exposures on Navajo are of No. 5 = Gray Cliff and perhaps there is some Vermilion Cliff (it may also be white).

A.T. 84° -- 23.189.

Bar. 21.44 at 3.40 pm

on point WB 57: DT 83.

Made camp on the plateau.

Wednesday, August 4. Camp 27 = 23.01 = 8120

Camp 25 = 24.64 = 6240

Bfst at 9 am after 6 miles—a portage.

{p. 53} **Thursday, August 5.** Camp 25 = 24.71 = 6180 to Camp 28

I ride through the hills W of Last Chance crossing the sag of the foot spur, and hunt fossils. *"Gryphaea"* abounds at the base of the Gray Shale but I do not find it unquestionably below. It probably however occurs also 30 ft below the base of the Gray under the yellow sandstone and at the lower coal horizon—here marked by a lignite or bituminous shale. I found some other obscure fossils below the Gray but the oyster is the only thing well preserved. It is now evident that the valley within the Pine

Cr. section is due to the gray shale, and T of that section represents not the lower Straight Cliff but the base of the upper cliff. The *"Gryphaea"* horizon is the same in each.

Loc. 913 is at the bottom of the gray shale, just above the sandstone in which it rests.

{p. 54}   Loc. 914 is in the gray shale below the middle.

Loc. 915 is the sandstone and shale below the gray shale.

All these are on Last Chance Cr.

Camp 28 is not far from Camp 21 and is on Pine Cr.

Friday, August 6. Camp 28 = 24.30 = 6635

The gypsum of the Jura which is so conspicuous here as well as all the way from Supply Camp to Last Chance does not show—that I could see—at the end of Kaiparowits.

After passing the site of Camp 20 we do not pass to the gray shale valley, but stick to the gypsum monoclinal valley up to where it heads under the trachyte debris. Then we turn to the left onto a ridge of Vagabond sandstone and barely touch the gray shale before leaving the seds. The vagabonds are at the last view nearly all sandstone; the conglomerates above are

{p. 55}   not over 50 ft thick and weather so as to abandon the crest. I saw nothing of the yellow sand at the base of the gray shale.

Saturday, August 7, 1875. Camp 29 = 22.70 = 8500.

Top of G.M. at the top of the 19–20 fold = 21.60 = 9860.

Camp 30 (21.61 = 9840) is between camps 18 and 19 and on the outlet of Lake Aspen.

Sunday, August 8. Camp 30 near Aspen Lake = 21.65 = 9790.

We cross Aquarius. The valley we strike in descending is structural. The opposite wall shows bedded lava (Fig. 2.38).

{p. 56}   The trend is a little E of N of the great fault on the opposite side which trends N.

In distant Rabbit Valley is a low ridge that looks like fresher lava than the general sheet.

Noon camp = 21.55 = 9920

Made Supply Camp at night

The Pine Creek section gives:

Straight series	2300
Gray shale	1300
Lower coal series	100
Vag. + Pothole	1000
Lower Jura	450
G.M.	1550
	6700

For 6 m. before reaching camp we could see at the right exposures of red white and yellow soft rocks that I suppose to be Jura. At the only point I examined them is a yellow conglomerate over red shale.

Figure 2.37. "Lazarus, Duke of York." This mule presumably was obtained from York Ranch. Gilbert reproduced the head in the first edition of his monograph, titled "Ways and Means". The illustration was omitted from the second edition. See comment in Introductory Chapter.

Figure 2.38. Wall of canyon descending from the Aquarius Plateau "shows bedded lava (a-a) with a max visible depth of 700 [*feet?*]."

{p. 60}   Altitudes corresponding to barometric inches—
unreduced

Inches	Feet
18	14,00
	140 for each 1/10 inch
19	12,600
	135 for each 1/10 inch
20	11,250
	130 for each 1/10 inch
21	9950
	125 for each 1/10 inch
22	8700
	120 for each 1/10 inch
23	7500
	115 for each 1/10 inch
24	6350
	115 for each 1/10 inch
25	5200
	110 for each 1/10 inch
26	4100

$1° = 1/57 = 90$ ft to the mile
$2 = 1/29 = 180$
$3 = 1/19 = 280$
$4 = 1/14 = 380$
$5 = 1/11 = 480$
$6 = 1/10$
$7 = 1/8$
$8 = 1/7$
$9 = 1/6$
$10 = 1/3$

Letter list
July 8 - No. 25 - Home
July 11 - No. 26 - Home
July 15 - No. 27 - Dutton
Aug. 9 - No. 28 - Home
Alt. Azimuth corrections.
    For angles read on inner scale 3/8°
                      outer scale 9/8°
    Determined July 23d.

No error of needle of alt. az.

[*A table of "Comparisons at 7 am and one 7 pm reading" is not explained and is ommitted here.*]
[*This notebook concludes with an unlabeled drawing of a scorpion, Figure 2.39.*]

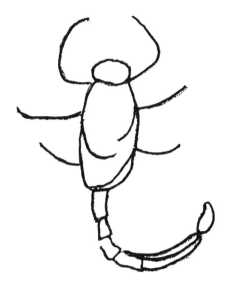

Figure 2.39. This unlabeled drawing of a scorpion is at the end of Notebook 2. One must imagine the incident that caused Gilbert to make the drawing.

CHAPTER 3*

Chapter 3. Notebook 3. Supply Camp, Aug. 10, 1875. Aug. 11 at Supply Camp, recognition that the lavas are faulted, climbing Aquarius. Aug. 12 to camp 31 in Sulphur Creek fork of Pleasant Creek northeast of Lion Mountain. Aug. 13 to Camp 32 in Sheets Gulch area. Aug. 14, Camp 32 held; exploring junction of the Waterpocket fold and east end of the Teasdale anticline. Aug. 15, camp 33 probably south of Sand Creek. Exploring junction of the two folds. Aug. 16, camp 33 south along the back of the Waterpocket fold to camp 34 near the Bitter Creek Divide. Aug. 17, continuing south along Hall Creek to the Red Slide at camp 35; water hauled to horses because canyon with water holes too narrow for the animals. Aug. 18 to camp 36 in Hall Creek (south of the Red Slide ?); climbing the Waterpocket for views south, and east. Aug. 19, camp 36 held; exploring Hall Creek's incised meanders and the Waterpocket fold. Aug. 20, march back north to camp 37 near The Post. Aug 21 to camp 38 on the southeast base of Mt. Hillers (Starr Spring?). Aug. 22. Examining southeast edge of Mt. Hillers. Aug. 23, with Graves, Adams, and Sorensson started for Mt. Ellsworth. 23 and 24 at Mt. Camp on Mt. Ellsworth.

*For abbreviations of geological terms and place names, see Table 0.1 and appendix.

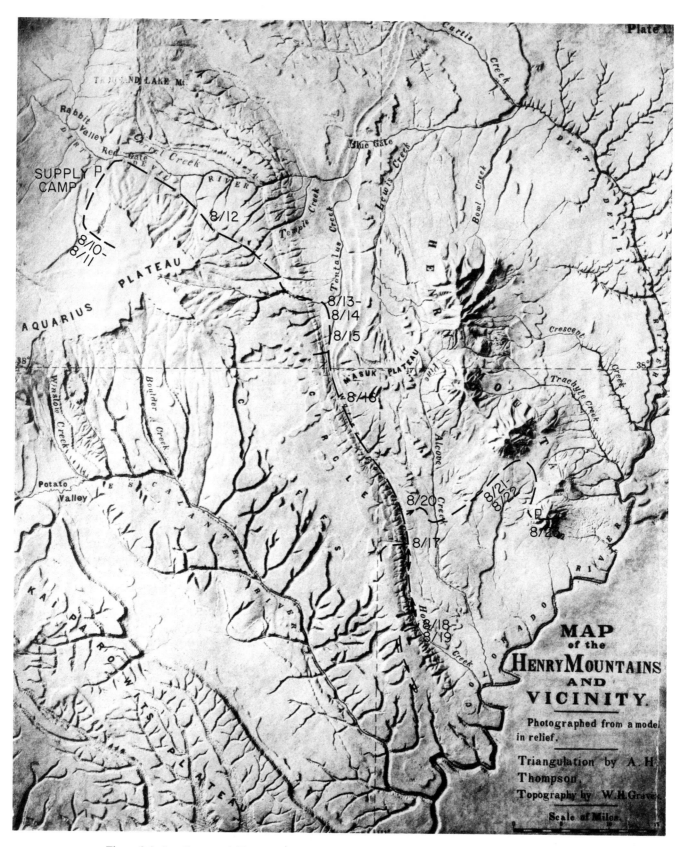

Figure 3.A. Landform map illustrating Gilbert's route from Supply Camp at Red Gate to Mt. Ellsworth, August 10 to 24, 1875. See also Fig. 0.D for route near Mt. Ellsworth.

Figure 3.B. Oblique airphoto of the south side of Mt. Hillers. Gilbert's route was east across the gravel covered pediments sloping southward from the mountain. This is the Hillers stock and the shatter zone around it; all the laccoliths on this mountain were injected northward and northeasterly from the stock. The white sandstones turned up almost vertically at the base of the mountain belong to the Glen Canyon Group plus some Entrada Sandstone (Gilbert's G.M., or Great Massive, and Vagabond). Permian formations border the south side of the stock but all these formations are cut off discordantly northward so that the wall rock outside the shatter zone on the north side is formed of Cretaceous formations. Left of Mt. Hillers can be seen the Waterpocket Fold, and on the skyline above it is the Aquarius Plateau (L) and Thousand Lake Mountain (R). Mt. Ellen is at the right of Mt. Hillers; Mt. Pennell projects above the Hillers summit. (Photograph by Fairchild Aerial Surveys.)

G.K. GILBERT, POWELL SURVEY, BOX 806,
WASHINGTON, D.C.
NOTEBOOK NO. 3
OPENED AT SUPPLY CAMP, UTAH ON AUGUST 10, 1875
CLOSED AT CAMP ON BASE OF MT. ELLSWORTH, AUGUST 23, 1875

{p. 1}   **Tuesday, August 10, 1875.** Supply Camp 23.31 = 7770

3 mi. N from Supply Camp the foothills of M.L. [*Thousand Lake*] show much disturbance. Standing on Y (see sketches p. 1 and 2 [*Figs. 3.1–3.4*]) I am on lava striking NW and dipping SW at 45°. I can trace the south arm across a creek to the NW.

There and here the lavas overlie or rest against a non-conforming series of seds. which dip 45° to the SSE.

{p. 2}   The lava section consists of bedded lava (a) in large part brecciated—100 ft. [*Some of these lettered units do not appear on the drawings.*]
b) ashen tuff about 100 ft at the point of measurement but varying with the uneven surface of sed.

The following is the section of the sed.
d) = pale arenaceous shale, perhaps in part tuff, with beds of limestone containing casts of shells (coll)—Ter ? 100 ft
E) = pale gray conglomerate (similar to that seen 10 mi further south) calcareous and alternating with cream shale (selenitiferous) 100 ft

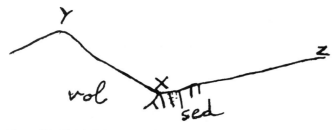

Figure 3.1. Unlabeled cross section on p. 1 of Notebook 3. Probably across the Thousand Lake fault near present town of Lyman.

$x = 22.93 = 8210$. Dip $= 45°$
x from $y = 21^{da}$. $y = 22.85 = 8310$. z from $y = 5°^{ea}$.

Figure 3.3. Unlabeled cross section described on p. 2. Probably crossing the Thousand Lake fault near the present town of Lyman.

z from $d = 12^{ea}$. $d = 22.91 = 8040$

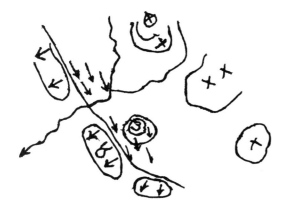

Figure 3.2. Unlabeled sketch map on p. 2, probably the Thousand Lake fault near present town of Lyman.

Figure 3.4. Continuation of section in Figure 3.3.

w from $z = 23^{da}$
v from $z = 10½^{da}$
u from $z = 1^{ea}$
$z = 22.79 – 8390$

{p. 3}   This passes into
f)   = pink and white shale and limestone   100 ft
g)   = red and white slaking aren. shale   50ft
        largely covered by debris from
h)   = calc sand, fossils, cream like that above the foss.
        Jura. The top of this is the summit z (p. 2).   300 ft
    In the level mass to the NE, the base of G.M. is on a level with z. The line of separation is an uneven one trending for ¾ of a mile NW by W. h is less 'fine' than usual and runs in part to shale. The lower 20 feet is red and decidedly arenaceous. Plenty of fossils found later.
i)   = massive white sandstone   30 ft
j)   = unseen and some shale   300 ft
        $w = 22.75 = 8140$
j2) = unseen   400 ft
k)   = yellow mass. sandstone and unseen   200 ft
        $v = 22.96 = 8170$ to v
j2 and k are G.M. Following the creek to its springs I find the main mass of G.M. [*Navajo Sandstone*] dipping to the SW and climbing out to the left I find myself in Jura (h). It is plain that the series of which I have made a section is a small block thrown askew

{p. 4}   on the line of the fault—one of a number.
[*Recognition of the displacement along this fault, the Thousand Lake fault, revealed to Gilbert that these displacements are middle Tertiary or younger, whereas the folding was earlier, as revealed by angular overlap of the early Tertiary formations. At that unconformity, Cretaceous formations are cut off between Salina Canyon and the Red Gate.*]
    From my farthest point it looks as though the fault trended NNW (3 miles) and S by SE (10 mi: the Button). These fragments give little idea of the whole throw which is equal to the altitude of Aquarius and M.L. above the Dish.
Wednesday, August 11. Supply Camp 22.19 = 7910
Foot of climb 20.63 = 11,080
Top of Aquarius 20.20 = 11,670
Saddle   = 11,370
Button = 20.15 = 11,750
    In climbing Aquarius I could see very plainly the fault along

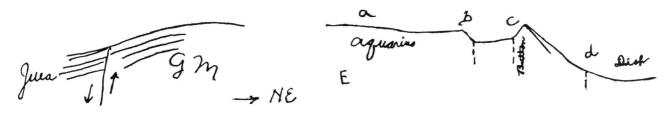

Figure 3.5. Fault in the G.M. at NE end of the section in Figure 3.4

Figure 3.6. Cross section at the Button.

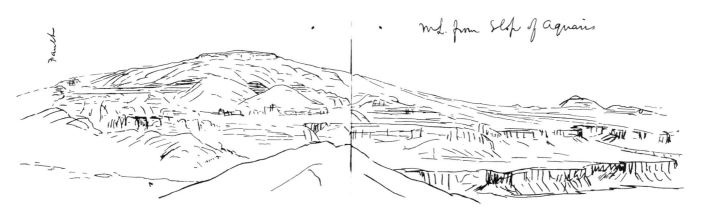

Figure 3.7. Sketch of M.L. (Thousand Lake Mountain) from north slope of the Aquarius Plateau. (See also Fig. 1E.)

the W side of M.L. [*Thousand Lake fault*] that I was on yesterday and I attempted a sketch to illustrate that and the faces of red rock (p. 5, Fig. 3.7).

The Button [= *Lookout Peak or the block north of it*] is not higher than the highest part of Aquarius but is higher than the nearest edge by 100 ft. (Fig. 3.6). However it cannot be a slide. It is set off ¾ of a mile from the plateau. There must be displacements at *b, c, d,* and probably at *a*. They can all be traced from here northward but not S. The Button ridge breaks down in a little distance to the N but reappears where we climbed Aquarius

{p. 6}   four miles back. A score of ponds are held by the Saddle. This set of displacements are wonderfully straight and parallel except *a* which is indefinite and may be no displacement at all. I note it because it may be the fading out of one of the lot first noted on Aquarius.

On getting a northward view (from a fir) I am disposed to think these displacements turn to the left at the edge of the plateau so as to strike for the M.L. fault and from it (Fig. 3.8).

{p. 7}   Just here this system of folds seems to be cut across by

a great displacement with S throw. —No. The facts will not warrant that. There is a break in the continuity. Then beyond the confusion and beyond our trail of Aug. 8, this Button ridge is resumed at a level 200 ft lower and continued on the same level for ten miles. It is limited by displacement *c* but *d* has disappeared (Figs. 3.9, 3.10, and 3.11).

{p. 8}   The mass of foothills that looked morainal Sunday still look so but the case is not strong. The lavas nearby are thin and rest on soft tuff and this might have given rise to a slide structure on a small scale. In the many lake basins contain not one lake. It is favorable to the moraine hypothesis that the wavelets of the tufa 'in' are arranged concentrically about the mouth of the big notch-like valley or gulch just south of the Button—thus (Fig. 3.12); and the area would naturally be larger if they were a slide system.

[*The geologic map accompanying U.S. Geological Survey Professional Paper 363 indicates these deposits are till.*]

The depth of lava on Aquarius is 500± ft.

The displacements in the Dish generally are small.

Thursday, August 12. Supply camp = 23.17 = 7930.

Figure 3.8. View north from the Aquarius Plateau across the Red Gate to the fault (f) along the west side of Thousand Lake Mountain (M.L.). (See also Fig. 1-E.)

Figure 3.11. Unlabeled map of Figure 3.10. "*b'* is in line with *b* but fades to nothing in 4 miles. *K* may be in line with *a* but probably has nothing to do with it. *k'* is local and fades both ways. *k* converges toward *c* then they diverge again and both weaken to the south. *m* is the fold with the satanic swell at its base. I cannot see how it terminates hitherward."

Figure 3.9. Map of the fault lines south of the Button (Lookout Peak).

Figure 3.12. Concentric arrangement of hillocks about the mouth of the big notch-like valley just south of the Button.

On the divide near Wheeler's Crag is the best illustration of dry sand erosion I have seen this year. The trachyte boulders are carved nearly as strongly as rocks at Fortification rock.

We dine at creek beyond W. Crag. Marches 7.45–10.45; 12.30–3.45 (tot. 6.15).

From the fold valley [*Sulphur Creek*] at □ Top [*Lion Mountain*] we

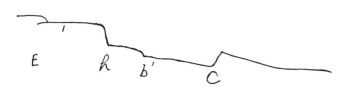

Figure 3.10. Unlabeled cross section on p. 7. Probably of faults along the west edge of the Aquarius Plateau near the top. Compare Figure 3.11.

{p. 9}   follow a mono. in Trias below the G.M. and chiefly below the Shinarump. At one point the cañon cuts to foss. lime

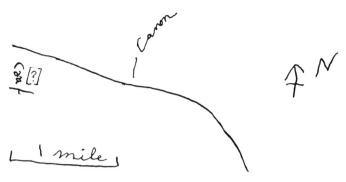

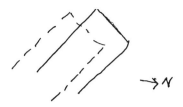

Figure 3.16. Ruled line shows the slickensides and the dashed lines the position of the next block east (p. 10).

Figure 3.13. This unlabeled drawing in the middle of p. 9 seems inconsistent with the text.

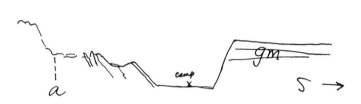

Figure 3.14. "General cross section at camp [*31*] and I suspect a fault at *a*" [compare Fig. 3.17].

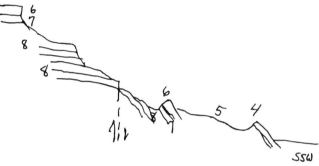

Figure 3.17. Cross section at fault suggested in Figure 3.14. "4 = Shin.; 5 = red shale and sand; 6 = yellow calc sand; 7 = red gyps shale; 8 = white sand unevenly bedded. Near the fault, 8 is shattered. Climbing to the top of 6 I find it dipping 3 degrees to the NNE."

Figure 3.15. Northward offset of the blocks described on p. 10.

{p. 10}   about vertical. The movements as indicated by slickensides were nearly horizontal but inclined 5° or 10° to the south. Going to the west, the successive blocks are set further to the north—thus (Fig. 3.15), and by this means a general strike of ESE instead of E is attained although each block dips S and strikes E. Looking at a block with the plane of the strike = plane of paper it is thus (Fig. 3.16). At one place the throw is 'variable'.

Farther up the hill I find that the red under the calc sandstone is gypsiferous and shaley, that the white below it is aren. and that there is a fault at *a* (p. 9) (Fig. 3.17).

{p. 11}   I can trace the band of the Shin. outcrop still further; in fact quite to the Escalante anticlinal. The pt. 119 is beyond that.

Top of 6 = 22.65 = 8670

The upper part of 8 is calcareous and full of chert and geodes. It equals the cherty limestone.

This fault is only local and terminates in 1½ mile each way.

From the top of 6 the base of the Shin. at the nearest point = 18°.

All south of the fold is a carved sea of G.M. with a half mantle of trachyte [*debris*].

that is probably Carb. We finally encamp above the Shinarump and I get on a crag near camp (31) to look about. The local dip is 43° to the S and the local strike is about ESE. The strike curves however and 2 miles away is much more southerly. This I note by tracing the Shin. which is white and conspicuous though only 30 ft thick. The beds above it are locally concealed.

Below are red sandy shale changing to shaley sands. There are several hundred [*feet? beds?*] and are followed by yellow sandstone (?) 50 ft. Then a little red again and then some pale beds that may be calcareous (Fig. 3.14).

Later I find the yellow rock so calcareous as to be near lime as sandstone. When I strike them they exhibit a curious faulting. The 'local' dip of blocks is to the S at 50°. A set of faults strike S and are

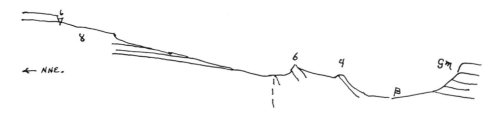

Figure 3.18. Section approximately to scale across the faulted SW end of Miners Mountain.

22.65 = 8670
6 = gray lime-sand weathering yellow
        22.65 = 8560
7 = gyps red shale sandy, 40 ft
8 = white calc. cherty geode sand changing at 8400 to white sand
        heavy bedded. This is visible to 23.13 = 7980 where it is
        covered by the upper part of No. 8 beyond the fault.
At the base of Shin. ridge = at camp 31   23.30 = 7780.

Where we struck fossils this pm the only shell I recognized
was a small *"Schizodus."* [*Another example of Gilbert using his
paleontology. This Permian fossil helped him determine the base
of the Triassic.*] The

{p. 12}   horizon was probably that of No. 6.

The base of level G.M. S of camp is somewhat below the
level of camp—perhaps 100–200 ft. On the lower half of this
page I have drawn as well to scale as I can by eye estimate the
section across the fold here. Making 6 and all below Carb. re-
quires that an outcrop in cañon of the last creek we crossed—just
W of Square Top butte [*Lion Mountain*]—be put down as Carb.
I did not visit it but the E wall of the creek cañon showed a white
series below the red of the sub-Shin. sands [= *Moenkopi Formation*] (Fig. 3.18).

The only water we saw after passing the Square Top divide
is half a mile back. It is a small stream [*Sulphur Cr.*] from G.M.
and sinks at once in the sand of the arroyo we followed from the
divide. The water we encamp on is evidently permanent but is a
small stream. It probably runs clear down to the cañon 2 m
ahead, where the drainage crosses the round top plateau [*Miners
Mountain, to join Pleasant Creek*].

{p. 13}   **Friday, August 13, 1875.** Camp 31 = 23.33 = 7740
= 7:15 to Camp 32.

At 23.60 = 7430 and in two miles we find the head of the
cross cañon and a big creek [*Pleasant Creek*] running through it.
In the cañon nothing is seen below No. 8. At the cañon there is no
fault but the DD fold has its normal character. Following its
trend eastward Graves and I first pass up a red wash [*Tantalus
Flat?*] with a monoclinal slim ridge dividing it and walled on one
side by G.M. and on the other by sub-Shin. Toward 119 is the
Shin cliff again dipping SE and the upper end of the red wash is

an anticlinal of which the trend is ESE while the whole fold
descends in that direction. Passing over the divide and descending
toward the next cross cañon [*Sheets Gulch*] we find the character
of the anti. changing. The steep dip to the south is greatly diminished but it still is slightly stronger than that to the N.

Standing at our farthest point (23.55 = 7470) [*head of
Sheets Gulch*] I am on red shale 500 feet below the Shin. NE the
Shin. is 500 ft above me and ¼ mile E it is on my level and ¾ mile
SE and S the same. SW

{p 14}   it is 200 ft above and ¼ mile W it is 300 ft above and
½ mile [*sic*]. The fold trends about SE and but for the stronger dip
(locally) at the right would be the Escalante alone.

Fine ripplemarks [*well-developed ripple marks are a characteristic feature of the Moenkopi Formation in southeastern Utah.*]
As a rule their acute crests are on top. Also cross rippling. On all
sides that I have noted Shin. the G.M. is not far behind it.

Dinner camp near mouth of red wash = 23.60 = 7430.

After dinner I find and collect *"Schizodus"* in No. 6. There
is also a smooth or radially marked shell that I take to be the
other valve.

Marches 7.15 to 9.15—2 to 6.

In the cañon we find the white Carb. sandstone [*Coconino*]
thoroughly massive and cross-laminated. Its cliffs are stained
brown red like those of the Redwall limestone and in this it is
distinguished from the Trias. and Cret.

The dip all through the cañon is surprisingly small ranging
from 3° to 6° by guess. Trachyte debris all the way. A cross valley
at the Shinarump where the conglomerate does not appear and
the var. shales [*Chinle Formation*] are conspicuous. The Gray
Cliff sand shows as thicker than the Vermilion.

'Shinom' and etchings on Verm. (Fig. 3.19).

From a gravel table (trachyte [*debris*]) near camp 32 I make
a

{p. 15}   distant slope of upper G.M. 7°. The gypsum horizon
of the Jura is well developed but not the fossiliferous. Above the
gypsum is a great thickness of red shale but perhaps not greatest.
It culminates in Cliff A (Fig. 3.20) which = Belted Cliff and = the
Pothole conglomerate and vagabond series. I think B is gray with
a yellow cap and shows sign of coal.

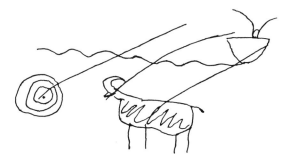

Figure 3.19. Etchings (petroglyphs) in canyon of Pleasant Creek. (These are crossed out in the notebook. Apparently there is another notebook not obtained by me with these and other drawings.)

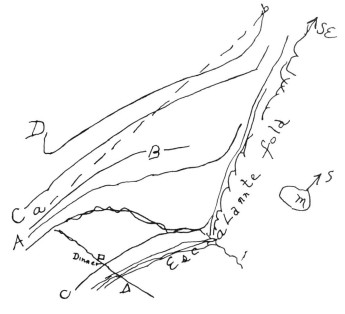

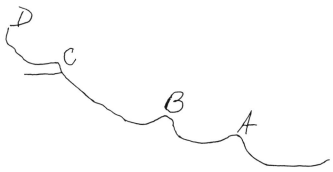

Figure 3.20. Profile of section summarized on p. 15.

Figure 3.21. Map showing beds A, B, C, and D along the Waterpocket Fold. (A = Dakota Sandstone = Gilbert's *"Gryphaea"* bed; B = Ferron Sandstone = Gilbert's Lower Sand; C = Emery Sandstone = Gilbert's Gate Sand; D = sandstone of Mesaverde Formation.)

C is the great gray shale ? with a yellow cap. [*Emery Sandstone*] D has the heaviest sand of all—yellow over gray. [*Mesaverde Formation*]

C makes the mesas to the N. [*Caineville Mesas = Twin Mesas, forming the Blue Gate*]

24.17 = 6780 = trachyte mesa near camp. It is a thin layer of debris preserving shales and there are others like it at the north. The slope of all is very gently eastward indicating that the source is Aquarius Plateau.

Camp 32 = 24.36 = 6570

{p. 16}   **Saturday, August 14.** Camp 32 = 24.41 = 6510 = 7 am

Dinner camp 24.49 = 6430 = 3 pm
Camp 33 = 23.94 = 7030
Marches 7.15–11 and 3–6.

The arroyo we followed to Pleasant Creek received water from three tributaries last night but is dry nevertheless. Near its mouth it is choked by sand. At Pleasant Creek where we first strike it, the water is 30 ft down in a deep cañon of dirt. On top = 24.65 = 6250. This is near its cañon through cliff A [*Dakota Sandstone*]. Dinner camp is on the wash from cañon 119. Graves and I climb on the Trias and Jura to 23.51 = 7440. The dip is to the E and 6°^{da}. The foss. Jura is here as strong as ever. I probably overlooked it yesterday. The gray cliff is 1000+ ft and the cañon through it is impassable. I think there is coal in B cliff [*Ferron*

Sandstone]. The gentle dip of the Escalante heads down to the DD fold. Beyond Pleasant Cr. its monoclinal dip is 45° at least and the strike turns 45° from S to SE. The way the cliffs turn is readily sketched on the next page (Fig. 3.21).

The throw of the Escalante must be 3000–5000 ft. The top of the G.M. on one side is higher than cliff D [*Mesaverde Formation*] opposite. Cliff D is beyond the influence of the fold unless at one point to

{p. 17}   the SE from here. Cliff C [*Emery Sandstone*] dips all the way until near the Island Mesa [*Tarantula Mesa*]. A and B dip all the way. There seems to be a max of dip along the line *a b* which may prove to be an independent fold.

Dinner Camp at 3 pm 24.49 = 6430. In the gypsum. Gray shale rather prevails with the gypsum but above to the top of the Belted Cliff, all is red except a band that may be the Vagabond and that sometimes appears on the face of the cliff and sometimes at its base or on the plain.

There are three lines of possible travel—1) at the base of the cliff, 2) at the mono above the gyps, and 3) at the mono below the gyps. Perhaps the crest of the gyps offers

{p. 18}   another way.

Approaching camp 33 (which is on waterpockets in upper G.M. a little SE of Pleasant Cr. [= *Sand Wash*]) I see on the farther side of Pleasant Cr. Vagabond [*Entrada Sandstone*] beds dipping (10°) to the SW, or against the dip of the Escalante fold. This can only be the DD monoclinal after crossing the Escalante fold. If a contour be run on the foss. Jura along the line of max flexure, or rather at the base of the hill, it will curve like the heavy line CD (Fig. 3.21).

I cannot see the DD fold cut the lettered cliffs [= *Cretaceous*] nor can I see it beyond them, but I think it affected the trends of these escarpments.

The dip is about 36° max of mono.

Sunday, August 15. Camp 33 = 24.03 = 6940 = 6 am. [*Sand Creek benches*]

Climbing Tit Mesa [*VA BM 7640, Notom Quadrangle, on JK at NE corner of Circle Cliffs*].

The local dip 100 ft above camp is 28°. It is less at camp.

At 22.93 = 8220 we are fairly on the plateau and have little if any dip. The horizon, 300 ft above the top of Vermilion [*my comma after horizon*]. The base of Vermilion must be as low as camp for the rock rises toward the hole in the wall.

{p. 19}   A level line cuts 200 ft below base of Verm. on 119.
Base of Tit Mesa = 22.85 = 8310
Top of Tit Mesa = 10.30 = 22.620 = 8600
Top of Tit Mesa = 12   = 22.610
Top of Tit Mesa =   2.30 = 22.520
Top of Tit Mesa   3   = 22.8300
Our travel on the Mesa was at the top of the Verm.

I see no good lithological reason for the preservation of this mesa on the very crest of the fold and of Circle Cliff. The cap sheet is somewhat calcareous.

No sign of water inside the Circle Cliff except some in Shin. cañons.

The N wall of the Circle dips a little to the N and makes one side of a gentle synclinal (Figs. 3.22 and 3.23).

Figure 3.22. "The north wall of the Circle [*Cliff*] dips a little to the N. and makes one side of a gentle synclinal *a* which runs parallel to the DD folds. It is very gentle and I cannot trace it into the Aquarius Plateau. Perhaps Tit Mesa [?] is a feature of it" (p. 19).

Figure 3.23. "At the first line of Shin. Islands the max. of the fold is 2 miles from the mono. At the Vermilion Island it is close to the mono., and at the large Shin. it has an intermediate position. This is as though a faint anti. ran obliquely along the Escalante fold—one which would trend toward Tantalus Point" (p. 20).

{p. 20}   Southward the E Circle wall has an E dip of 5° at first and increasing as the wall approaches the mono.

Cliff D [*Mesaverde Formation on Tarantula Mesa*] is the highest visible on this side Henry Mts. and its step is as great as that on Cliff C [= *Emery Sandstone*]. Its slope is characterized by interrupting bands while that of Cliff C is not. Its table rises with an increasing dip toward the saddle bet. H1 [*Mt. Ellen*] and Pennell and can be traced far toward the divide. The chief face of the cliff is W but it faces S also and its most southerly point is not beyond the head of W.-P. Cañon. Cliff C is a hogback all along the W face of D but makes another step 5 m further S where it swings E from the W.-P. Cañon [= *valley of Hall Creek*] it is controlled by a fold with NE throw; i.e., the salients on the SW edge of the C table are turned up as by a mono. Perhaps the cross fold that influences the line of max elevation of the Escalante runs there.

From the N end of the Table—I can see on the farther wall of a cañon that the Verm. is bent 20° without fracture (Fig. 3.24). Nearby it is bent more abruptly but I have not the same opportunity to trace continuity.

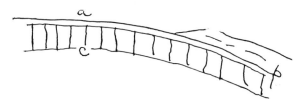

Figure 3.24. "I can see on the farther wall of a cañon that the Verm. is bent 20° without fracture [*p. 20*]. If *ac* is 4,000 ft, then *cb* is 5,000 ft."

{p. 21}   The NW face of D Cliff is the limit of the influence of the DD fold. The lettered cliffs curve around toward the Escalante mono and when they strike it turn abruptly—at right angles almost to the south and fall into the Escalante line. Nor does any sign of the displacement they have been following and revealing show itself beyond the Escalante mono. The contrast between Figures 3.26 and 3.27; 3.28 and 3.29 illustrate this.

{p. 22}   The plateau D (Figs. 3.25, 3.27, 3.28) is the lowest piece of country structurally in all this region, just as the Round Top Mesa is the highest.

The Belted Cliff is locally broken up and subdivided. The section of the G.M. is typical—a = white massive; b = less massive or massive bedded and banded in color; s = red massive—undermined into columns by the shale below (Fig. 3.30).

Descending on the fold to 22.98 = 8150 I find that I am still higher by several hundred feet than the D. Plateau.

The Escalante mono turns to a more southerly course (5°) where the lettered cliffs join it.

On the way. I am on top of Cliff D at about 23.60 = 7440.

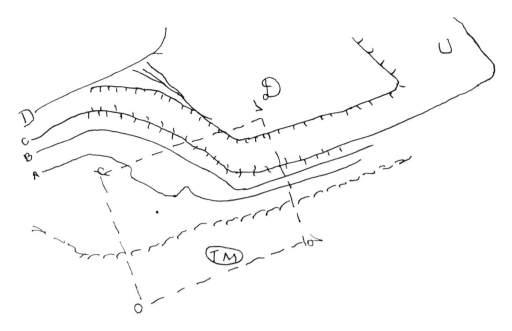

Figure 3.25. Map of the escarpments shown by the cross sections in Figures 3.26, 3.27, 3.28, and 3.29.

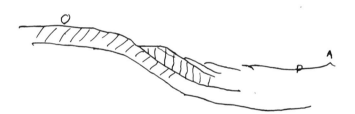

Figure 3.26. Cross section of the escarpments shown in Figure 3.25.

Figure 3.29. Cross section of the escarpments shown in Figure 3.25.

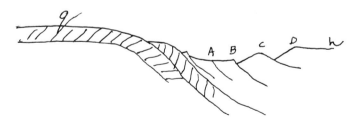

Figure 3.27. Cross section of the escarpments shown in Figure 3.25.

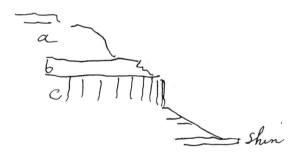

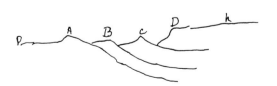

Figure 3.28. Cross section of the escarpments shown in Figure 3.25.

Figure 3.30. "The Belted Cliff [*Morrison Formation*] is locally broken up and subdivided. The section of the G.M. [*Great Massive* = Navajo Sandstone] is typical. *a* is white, massive; *b* is less massive or massive bedded and banded in color; *c* is red massive—undermined into columns by the shale below" (p. 22).

{p. 23}   **Monday, August 16.** Camp 33, 8.25, to Camp 34, 12.25.

At camp 33 the following section [*lower part Carmel Formation*] ascends from the G.M. white—

a)	yellow massive sand	20 ft
b)	red shaley sand	12 ft
c)	white heavy bedded to massive sand	20 ft
d)	red sandy shale	8 ft
e)	olive calcareous shale (foss)	15 ft
f)	red shaley sand	10 ft
g)	slaty cream lime sand	20 ft

In e are *"Camptonectus"* and *"Ostrea"*— This changes into white gray

h) shale above which alternates with gypsiferous for 150–250 ft

Camp 33 = 24.07 = 6900

All the cañons that head west of the Escalante anti and cross it bring down trachyte. All which head in it and cut deeply (there are 3 between camp 33 and Tit Mesa) bring foss. wood of the Shinarump.

The Belted Cliff beds are about the same as on Pine Creek and I even found a conglomerate near the top. There are however some soft high-tinted beds under the yellow sandstone at the top of the series, and in that sandstone I find *"Exogyra costata"* and *"Gryphaea Pitchnia?."* This bed is entirely below the B cliff and does not everywhere hold as escarpment.

{p. 24}   It is a back ridge of cliff A [= Dakota Sandstone]. Camp 34   24.20 = 6740 = 2.15 pm.   z (Fig. 3.31) = 24.00 = 7020 = 2.35

The yellow bed at top of G.M. is here calcareous and ferruginous and hardened so as to hold a line of hogbacks 2–300 ft high. The local erosion is strictly consequent—cataclinal from the E fold with monos between the catas (Fig. 3.31).

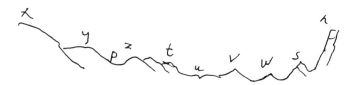

Figure 3.31. Profile and section at Camp 34 (marked by flag between y and z). "1 = Vermil. sand culminating in top of E. fold = x; 2 = Gray Cliff sand white mass; 3 = yellow = yellow brown calc ferr sand forming z—12 ft; Dip 36° to the NE. Strike NW by N. Direction of section NE by E." The numbers 1 through 5 are not shown on Figure 3.31; 1–3 are on Fig. 3-34; 7–11 appear on Figure 3.32, which seems to be a continuation of this section.

In Figure 3.31, the strike changes at once to NW. In the opposite direction it is SE by S and changes to SE by E.

From Z.   x = 10 1/2^{ea}; t = 15da; u = 13 1/2da; v = 4 1/2da; w = 4da; S = 1 1/2da; r = 2^{ea}.

4	= red sandy and shaley Jura (white at base)	50 ft
5	= gypsum and shale, white and red	150 ft
	t is 20 ft higher than summit of 5	
t	= 24.22 = 6730 = 3.15	
u	= 24.29 = 6640	

{p. 25}   from u–t = 9 1/2^{ea}
          v = 11 1/2^{ea}

6	= pale red and white sand soft as shale	500 ft ?
7	= cherty gray limestone. The chert is conspicuous and made of chalcedony and jasper. This constitutes v.	10 ft
v	= 24.12 = 6840 = 3.40. Dip 30°?	
	From v, r = 4^{ea}	
8	= conglomerate and shale, white and red	50 ft

Thence to W the rock is seen only in 3 ridges (Fig. 3.32).

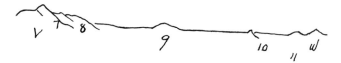

Figure 3.32. Continuation of cross section in Figure 3.31.

9	= a ridge of yellow and purple conglomerate to be accounted with 8
10	= a low ridge of yellow sand with many fossils. *"Ostrea," "Gryphaea"* (Loc. 917), *"Exogyra,"* and several other genera.
10	= 24.22 = 6720 = 4.40
11	= yellow shale with bands of sand and yellow and gray shale continue to w.

12	= w = yell and white sand heavy bedded	50 ft
	dip = 50°	

from w s = 4^{ea}
       r = 7^{ea}
       s^1 = 6 1/2^{ea}
w = 24.23 = 6715
s^1 = 24.27 = 6670 = 5 pm
from s^1, s = 7^{ea}; r = 8^{ea}

{p. 26}   From w to s (Fig. 3.33) are 13 and 14.

Figure 3.33. Continuation of cross section in Figure 3.32.

s = 24.06 = 6910 = 5.10 dip = 23° strike = SSE

13	= gray shale weathers yellow	
14	= yellow sandstone	100– 500 ft
14ᵃ	= same	
15	= pale shale with dark band and lignite,	30 ft
16	= white-yellow sand and sandy shale with dark streaks but no coal. About 300 ft	
17	= gray shale with slight interruption to 23.98 = 6099. From 15 to 17 (top) the dip has been steadily diminishing and now it is 7°.	
18	= gray shale and yellow sand alternating in the beds 23.87 = 7110	
19	= sand yell. mass	20 ft
20	= unseen—prob shaley sand	50 ft
21	= sandstone—yellow massive (shark's teeth) and the 'spring stem'	200 ft
r	= top of [*Station*] 21 = 23.64 = 7380 = 6 pm	
	dip = 7°	

from r.   x = 2 3/4ᵉᵃ
           s = 11 1/2ᵈᵃ

The dip ends in ½ mile and then begins the rise (dip) toward the Henry Saddle. It is a small dip and at the base of the Mt. seems to increase.

All features of the fold

{p. 27}   continue south except that the Verm. as well as the Gray Sand becomes a hogback. The next salient of r = D cuts off much [*of the view*] and the NE view is cut off in the same way.

Tuesday, August 17. From camp 34 at Sorenson's pockets (24.21 = 6730) to camp 35.

Marches: 7.15–11.40; 2.30–5.30.

We cross the divide and strike a trail in the mono above the oyster bed [*Dakota Sandstone*]. The line of fold seems very straight just here but 3 or 4 miles below the summit turns first to the left and then more 'deaded E' to the right. For 2 or 3 miles the crest of the Gray Cliff sand is wonderfully level, like the old bed of Pine Cr. and this although the rock must dip 20° and is 1000 ft above base of erosion on this side.

The limits of the valley are the G.M. and Cliff C [*Emery Sandstone*]. Cliff B [*Ferron Sandstone*] is a simple and low hogback and the Jura and Cliff A [*Dakota Sandstone*] are an uneven mass of hogbacks.

Black shale in the oyster beds.

A thin *"Inoceramus"* just above Cliff B.

{p. 28}   Just between the turn in the C Cliff the dip diminishes on the fold. It has been about 30° all the way but there it is 15°. Beyond, opposite the Circle Cliff Island it increases again to about 30°. The decrease in dip is marked by a turn.

Dinner = 24.89 = 5960.

The last notes were taken close to dinner camp and opposite the bend of Cliff C. The Jura at dinner camp dips 21° and toward NW.

The xlam [= *cross lamination?*] of G.M. dips SE.

The bed resting on the G.M. is dark red, a shaley sand.

The Vagabond series [*Entrada Sandstone*] appear hard, and liver red at top and purplish 'tone'.

The gulch in which we find water is Pa-runa-weap [*Indian word = Roaring Water*] of the narrowest type. The horses can go only within 200 ft of the water on account of narrowness and we have to carry it to them. The walls rise about 400 ft and the gulch ends abruptly [*probably the first or second side canyon above the Red Slide in Hall Creek*].

Camp 35 = 25.15 = 5690.

The Vaga. series are a massive red sandstone several hundred feet thick of rather duller color than the Vermilion and of rounded contours. Round corners are everywhere characteristic of the harder beds of this group.

Just below camp a great

{p. 29}   talus on the right runs clear across the cañon. It seems to come from the base of S Circle Cliff and the mono side of the Islands of Verm.

The Belted Cliff series is hard—red and white and a conglomerate series just as under Kaiparowits.

Water rises in the cañon just below camp and sinks at once. It is perhaps brought up by the clay of the talus dam.

Wednesday, August 18. Camp 35 in Waterpocket Cañon = 25.33 = 5490.

Marches 7.15—

Above point on yesterday's march the arroyo cuts into the G.M. making a Horseshoe Cañon.

At a point again on today's the arroyo does the same only more so. We follow in for two miles and are blocked by quicksand. The walls are of the G.M. and attain 1000 ft (guess) in height. Returning we cut across the neck and encamp ½ mile below the end of the horseshoe. It's a strange watercourse, consequent (following a mono) for most of its course but inconsequent in two places and, according to Jack, in another at its mouth.

{p. 30}   Camp 36 is Waterpocket Cañon below the U = 25.77 = 5030.

The dip at the horseshoe is small falling as low as 15°–20°. At camp 36 it is 20° in lower Vaga. or gypsum (Fig. 3.34).

{p. 31}   We are at an upturned undulation of the Escalante fold. In one direction it runs S by SE 3 miles and then swings to SE or SE by E for 5 miles. For that distance the max dip seems to be 20° and uniform and beyond it cannot be seen. It either turns short to the right or stops.

In the opposite direction it runs NW 3 miles with a dip of 20° or even up and then swings to N by NE with an increase of dip to 30° or 40°.

It is as though it was crossed obliquely by a fold of some sort but no such fold is betrayed to the E. Still there is a low angle swell running eastward and fading out, a swell that may be anti though I see it only as mono.

Figure 3.34. Section across Waterpocket Fold and Halls Creek near camp 36. "1 = red, shaly and bedded sandstone capped by red and gray shale—50 ft. Dip = 20°. t from a = 17½^{ea}. from a = 13^{ea}; 2 = red and gray sandy shale, (no gypsum seen) 100? b is at its summit. t from b = 21¾^{ea} (a and b are on the level of camp); 3 = Massive soft red sandstone (Vagabond) – 400 ft. C is in 3 dip = 19°. t from c = 19½^{ea}. c = 25.80 = 5000. d is in 3 dip = 19°. t from d = 35da. e = 25.45 = 5370. e = top of 3 dip = 5° (?). t from e = 85°; 4 = sandstone like e rapidly alternating with chocolate shale—a ribbon series—75 ft. f = top of 4 = 25.40 = 5430. Dip = 15°. t from f = 90°; 5 = red, (purplish) and white, heavy bedded sandstone and conglomerate with white pebbles. t = 25.23 = 5600, dip = 20°. (p. 31) t is not the summit of No. 5 I can see 60 ft more of it. The dip carries the top bed 1–200 feet lower than this station before it flattens."

{p. 32}   It runs toward Ellsworth.

The mesa that is spread before us to the E, SE, and NE and which descends from the left with the last swell and from the right with the Escalante dip and which terminates in a few miles is of the conglomerate possibly capped—(yes) probably by the oyster bed over a small area on Graves Point No. 331 and over a larger including his point No. 333. Purple and gray shale shows below the bed I take for oysters.

Beyond the table for many miles the country is monotonously red.

Ellsworth shows no volcanic colors but looks as though built of the valley rocks (Fig. 3.36). In the region a I can make out no dip but in the regions b and C (Fig. 3.36) I measure dips of about 25°. It is another Navajo Mountain without the lava (?). One mile high. [Striking out this note provides an insight into Gilbert's thinking.]

Thursday, August 19. Camp 36 = 27.775 = 5020

The southern table made by Cliff C [Emery Sandstone] has a dip N and W and I see now no sign of a cross fold. I trace the bed to the flank of Hillers and do not see that its dip increases that way—a mistake probably. The table I wrote about is more likely to prove one of Cliff B.

{p. 34}   Climbing to the edge of the fold I see Ellsworth better. On this flank the dip is this way unmistakably. I can see the successive circling around it—red at base, then white—and the white probably caps the summit. The white underlies the red and the red underlies Cliff A. Hence the white is upper G.M.

Edge of fold 23.58 = 7450

The gorges of Ellsworth show no red below the white, but the Vermilion may be there bleached. The Vagabond red runs into Ellsworth and Henry V [Mt. Holmes] and the saddle between with even dip as though the two points might be a geological unit.

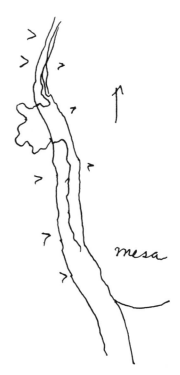

Figure 3.35. Map without label or explanation on p. 31. It clearly is a map of Halls Creek where it is superimposed onto the flank of the Waterpocket Fold, as described on p. 29 of the notebook. The "Mesa" is the bench capped by Dakota Sandstone.

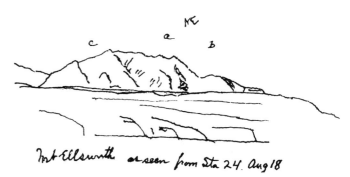

Figure 3.36. Mt. Ellsworth as seen from Sta. 24, August 18.

Summit—Geod. Pt. No XI
     1.30 pm = 23.43 = 7620.   2 = 23.46   4 = 23.42
4.30 = 23.29

The Escalante mono after swinging to the left (southward sense) as noted yesterday does not stop but turns to the right again and extends at least to the Colorado. The last course I can note is about S or S by SE. The crest falls to the S at first quickly and then more slowly, yet the crest is all the

{p. 37}   way composed of the upper white G.M. and far toward the Colorado there is even preserved (W of the crest) an island of lower Jura. Wherever I can see it the max angle of the mono is small, not exceeding 15°.

Figure 3.37. Waterpocket Canyon and the Horseshoe bend of Hoxie [= Halls] Creek as drawn in Gilbert's notebook (above) and as retouched by an artist and published in his monograph as Figure 68.

From this point to the Kaip. Plat. the dip is bet. 1° and 2°. The cross section farther S shows a still gentler western dip. It is retained, however, to some degree to the Colorado.

To the N the profile of the fold is peculiarly angular (Fig. 3.39).

Hillers must be trachyte-capped and trachyte around except its lowest flanks. On this side a half dozen spurs show fragments of red rock as sketched in Figure 3.40. This may be due to dip toward me and details of character look like that.

Looking N across cañon in the foreground I see in the G.M. the thickest single bed of sandstone I ever have (Fig. 3.41).

The oyster bench top of Cliff A follows the cliff down to about camp 35 and then runs across toward Mt. Ellsworth joining in to the area indicated yesterday. So there is a regular series

of steps from Cliff D [*Mesaverde Formation on Tarantula Mesa*] to the Colorado, the last of importance being Cliff A. Cliff A has here very much the character that it shows at the S end of Kaiparowits.

I am now doubtful whether the Horseshoe of Waterpocket is inconsequent. It occurs (see p. 31) where the mono flattens and may mark out an old mono valley when the base of erosion was 1000 ft higher. It is not without the range of probability that a stream in the progress of denudation should in one part cut straight down without regard to hardness, and in another follow the slopes of a hard rock.

{p. 39} **Friday, August 20.** Camp 36.—the most southerly point on Waterpocket = 25.90 = 4900.

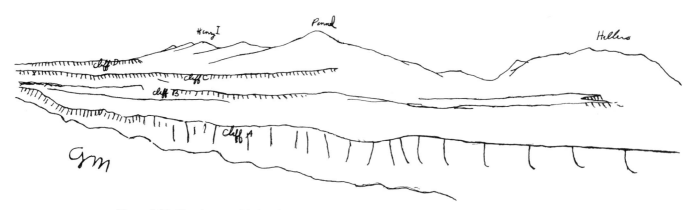

Figure 3.38. Sketch on p. 36 showing Mt. Ellen (Henry I), Pennell, and Hillers and the cliffs of the Cretaceous sandstones: A = Dakota Sandstone; B = Ferron Sandstone; C = Gate Sandstone; and D = Mesaverde Sandstone.

Last night Graves, Bell, and Sorenson and I slept out. We did not reach Point X1 until 1:30 pm and left at 5 pm. Darkness overtook us and we barely made a water pocket on the descent where we were forced by the uncertainty of the way and by weariness to stop. At 4:15 this morning we started again and reached breakfast at 6:30.

Hillers separated his outfit yesterday and, though he will travel and camp with us today, is now independent.

If the rock series had been classified in Utah, Colorado, Arizona, and New Mexico instead of in Europe, it would have taken some such form as the following.

    (a Freshwater = Tertiary

A

    (b marine (yellow sand and gray shale) = Cretaceous
    (c Belted Cliff series
    (d Foss. Jura

B

    (e G.M.
    (f Shin. series
    (g Carb.

C

    (h Sil.

D   = Archaean

{p. 40}   Summit opp. Horseshoe   25.58 = 5230
Creek bed above Horseshoe   25.65 = 5150

'I est.' the wash falls 250 feet in passing Horseshoe bend and cuts 800 feet into the sandstone. It would have to cut but 80 ft to pass along the soft rocks.

    Marches, 8.45–11.45, 2.15–5

Camp 37 = 25.00 = 5850 is a little south of our noon camp 3 days ago. It is based on a pocket the margin of which is grown with bullrushes and canes. The stream in the cañon from the head of the Horseshoe down is apparently permanent though perhaps not with the present size. It abounds in fishes—small but adult.

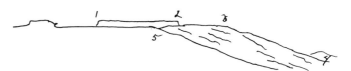

Figure 3.39. Angular profile of the Waterpocket Fold referred to on p. 37. "1–2 is the Vermilion Island inside Circle Cliff; 5–6 is the slope we climbed; 3 is the angle of Circle Cliff and 3–4 is drawn with slope a trifle too gentle; 3–5 is too steep."

Figure 3.40. South side of Mt. Hillers, described on p. 37.

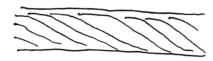

Figure 3.41. Bed of cross laminated sandstone in the Waterpocket Fold (Gilbert, p. 38). "The cross laminations make its individuality and it can be traced a thousand feet. From this point it subtends an angle of 1¼° and when Graves has plotted its distance I can tell the thickness. Meantime I estimate it 100 ft.   3° 8'
                                  4° 33'
                                  1° 25'
More exactly it measures 1° 25'. (Distance ⅞ mile; thickness 105 ft)".

Figure 3.42. "Ellsworth from pt near camp" [*Point No. 37*].

Figure 3.44. Beds rising onto the SW flank of Mt. Pennell as noted by Gilbert on p. 44.

Seeing Ellsworth from a new point of view I have the same quaquaversal and 'aniasiv' dips—at the left 30°; at the right 17°. See p. 41 (Fig. 3.42).

Pennell and Hillers still look very volcanic.

{p. 43}   **Saturday, August 21.** Camp 37 at Adams' water-pocket = 25.14 = 5200

Marches 7.10–11.30, 12–4

Passing through a gap in Cliff A we follow a mono at the base of Cliff C. At the highest point of the mono (24.80 = 6084) there is considerable dip 'to' Cliff B (say 10°) in sympathy with Escalante fold. At the top of B and on its face are thin coal seams—none economic. Gypsum gleams on all the gray shale slopes.

A mile farther the B Cliff becomes nearly level, its faint dip being to the N.

The big *"Inoceramus"* is in Cliff C shale.

From a point on the trail (Sta 4), I have a good view of Pennell. Cliff C stands nearly level at its foot and then curves up against its flanks.

The crags against Hillers that I noted as red [*p. 37*] from Pt. X1 I can now see are nearly vertical (see Fig. 3B).

Pine Alcove Cr. 11 am Lunch Cr.   25.10 = 5740

{p. 44}   From a hill near Hillers I can see the C Cliff [*Emery Sandstone*] rising against Pennell with a slope of 5° (Fig. 3.44), but I cannot trace it beyond (a). I can however see similar sandstone and shale in unsympathetic mass that probably are slidden and among them are similar masses of trachyte the position of which I refer to the same cause.

Camp 38 is on the SE base of Hillers. We have found several minute springs in skirting the mountain and this one barely suffices us. It cannot be depended on as permanent.

The rock which rises toward Hillers from the south is the B Cliff [*Ferron Sandstone*]. It is lost in the 'debris' without increasing the dip (7°) with which it approaches (see p. 45). But beyond are red and white sands—inferior rocks tilted almost to the vertical and interspersed with dikes. Moreover these sandstone hogbacks seem to trend in a curve around the mountain as far as they extend.

At Camp 38 = 24.00 = 6970 I am below the horizon of the B Cliff sand (Fig. 3.45). Broadly speaking the C, B, and A cliff rocks are synclinal from the Escalante fold to the Henry Mts, and so far north as [*?*] Pennell at least have a southerly dip. Beyond the Escalante mono there is a southward dip. Equally general and gentle all the way from the Circle Cliff Island to the Colorado.

{p. 46}   Hillers trachyte is a pale gray paste with large white xx [*crystals*] of feldspar and xx large and small of hornblende. This describes the variety at hand.

Figure 3.43. South end of the Circle Cliffs. "Escalante fold from near camp 37, August 20." View is from the east rim of Hall Creek = Gilbert's Hoxie Creek, about 8 miles north of the south rim of the Circle Cliffs.

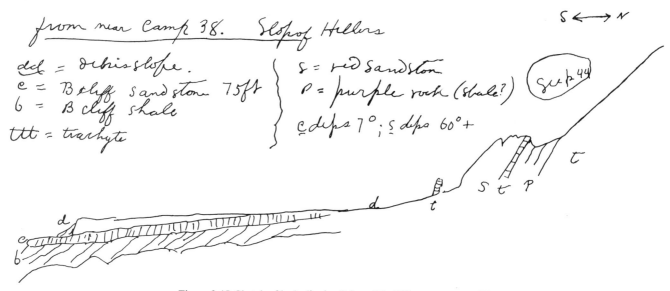

Figure 3.45. Sketch of beds dipping S from Mt. Hillers near camp 38.

Sunday, August 22. Camp 38 = 24.11 = 6850 (Fig. 3.46).

At the gate of the cañon *c* [*Not shown in Figure 3.46*] I read 23.68 = 7330.

Inside it and outside are red sandstones that I take to be Vagabond. A contact

Figure 3.46. "Camp from *a* = SSE = 6da. *a* = 28.875 = 7100. The rock at *a* dips unevenly from 0°–10°. It is sandstone and conglomerate of Cliff A" [*Dakota Sandstone = Gilbert's* "Gryphaea" *bed*].

"Across an arroyo to the W. the same (?) rocks lies thus—"

"The next exposure (b) does not appear directly on the line of section but at both 'sides'. At the west where I examine it, the rock is white sandstone and conglomerate (100 ft +). Dips 85° and trends W. *b* = 23.75 = 7250."

"There is then a hiatus of several hundred feet 'succeeded by' a greater distance of trachyte in dyke form."

{p. 47}   between dike and Vaga. shows little alteration of either. The trachyte shows almost none. The sandstone is bleached from red to green and yellow (pale) for about five feet. I call this locality of contact C in the [*specimens*] labels. The beds of trachyte and sandstone are each several hundred feet thick.

Most of the larger trachyte masses have their larger axes determined by the bedding but there are not wanting dikes which run obliquely. [*In written communication, David D. Pollard reports efforts to distinguish tilted sills here from dikes by paleomagnetism.*] There is considerable uniformity of dip and trend of the seds, 80°, 85° and ('here') ENE.

The dikes interfere with the measurements of the seds. I "'put' up" as follows—

The exposures at *a* and *b* (p. 46) are Belted Cliff. I can trace them over a mile as foothills parallel to the mountain front. Then come

400 ft of red sandstone (Vaga.).
150 ft of softer red and gray (Jura)
1000 ft of white and pale red sandstone (G.M.)
Then trachyte without interruption.

At *Loc. D.* A red sandstone is hardened at the contact but retains its color much nearer than at *C*. The whitened belt ranges from 2 inches to 1 ft.

{p. 48}   At the loc. the trachyte has a dark paste and is too much decomposed at contact to collect. Its mineralogic character seems unchanged up to the contact; but that surface is one of percolation.

If my identification of strata is correct, neither the gyps nor the sectile Jura are present. Still I can do no better.

Figure 3.47. "On this Mountain [*Hillers*] I see no strata 'curving' up. Not so on Ellsworth. For 4 miles, from its westward syn., the curves of bedding gradually increase until they are lost in the trachyte—for Ellsworth abounds in trachyte and much that I have taken for strata on its sides is trachyte in dykes [*sills*]. The red beds at the base are evidently Vag. [*Entrada Sandstone*]. So too the rounded hill with NW dip bet. Ells. and H.V." [*Notebook 3, p. 48*]. [*Actually, there is a substantial rise of the strata toward Hillers but the dip is obscured by the gravel on the pediments.*]

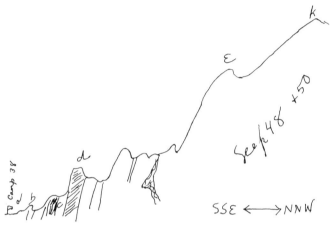

Figure 3.48. Cross section at the south base of Mt. Hillers, referred to on pages 48 and 50.

d is on trachyte (see p. 49). *d* = 23.20 = 7900. From *d*

    Camp 38 = 13½^{da} S by SE

    a = 21½^{da}

    b = 33da

    c = 50da

    e = 32 1/3^{ea}

    k = 32 1/3^{ea}

    e from camp = 20½^{ea}

    k from camp = 21 2/3^{ea}

Beyond where I can see seds of this outcrop I can see dikes of the same trend both E and W. The bending or swinging of the hogbacks about the mountain base is slight but unmistakable.

On this Mt. I see no strata curving up (Fig. 3.47).

The rocks that make up the slopes of H.V. [*Mt. Holmes*] are G.M. running continuously into Verm. cliffs toward the E. As far as I can see around the E flank of Ells. it is quaquaversal. On every side visible, H.V. is Belted [*not quite so and Gilbert later corrects this*].

The Leaden rocks [*Mancos Shale*] are limited southward by a line nearer Hillers than Ellsworth and running to the divide

Figure 3.49. Fossiliferous Cretaceous rocks referred to on p. 50.

Figure 3.50. "Henry V [*Mt. Holmes*] from camp 38. H.V. is the lowest of all, little beside its dikes remaining of the flow. Looking at it from a hill E of camp 38, I am impressed with the idea that the dikes are radial, diminishing outward. The dip of the sandstones is not greater at the center than on the flanks. It is just a tumor cracked in the middle. A main crack (dike) runs S (a-b), another NE (a-c), a third W (a-d). The S and W dikes show no flows and the adjacent sandstones are preserved by their hardness. Nothing below G.M. [*Navajo Sandstone*] shows but G.M. goes to the top" [*see p. 52*]

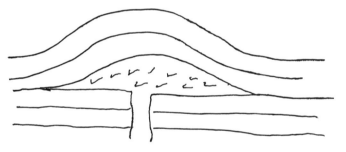

Figure 3.51. This now classical figure, which was later (Notebook 8, p. 48) described as a laccolite, appears without a caption on p. 51.

{p. 50}   bet. Hillers and H.V. My view through that divide reveals no Cret. and no Belted Cliff even.

On the descent I see a larger exposure of the *b* hogback (Fig. 3.48) and find it to consist of white and gray sand alternating with red shale. It must be the Belted Cliff cap. The *a* (Fig. 3.49) rocks are faced for some distance by a conglomerate of coarse pebbles (*a'*) set on edge and more too. In *a* I find Cret. foss.—"*Gryphaea*" indicating it to be the yellow sand above *b*.

After dinner (Fig. 3.50).

The types of Hillers and H.V. are somewhat different. The radial dikes of the latter are feebly represented by the thin radials of the former and the concentric dikes of Hillers do not appear in H.V. If Hillers be one extension of the H.V. type, then only the Trias was lifted and the Carboniferous either lay below the seat of action or below a distributing reservoir (Figs. 3.51 and 3.52).

Figure 3.52. Henry V (Mt. Holmes, smallest of the five Henry Mountains) as sketched by Gilbert from near the southeast corner of Mt. Hillers, August 22, 1875.

Figure 3.C. Mount Holmes, Gilbert's H.V., from a little closer and a little south of where Gilbert made his drawing (Fig. 3.52).

{p. 53}   **Monday, August 23.** Camp 38.   23.97 = 7000 = 6.10

With Graves, Adams, and Sorenson started for Ellsworth at 6.25. Mountain camp at 9.30 = 23.75 = 7250.

In coming we descend geologically upon strata that rise toward Ells. At first the dip is slight through the lower Lead [*Mancos Shale*] and Belted rocks [*Morrison Formation*]. In the Vaga. [*Entrada Sandstone*] the dip increases to 10° and 20° and in the G.M. [*Navajo Sandstone*] to 50°—a maximum. Then dip diminishes toward the mountain (Fig. 3.53).

Our camp is just below [*downhill*] the Shinarump and the Shin. dips 45°. The var. [*Chinle Formation*] shales are distinctly developed but I see nothing of the chocolate [*Moenkopi Formation*] below the Shin.

My identification on Hillers must have been correct.

Having climbed to 23.50 = 7550, I see a few hundred feet of Chocolate separated from Shin. by 300 ft. of trachyte. A division in the trachyte indicates that it was injected at two times.

{p. 54}   The lava is essentially the same as on Hillers. I see no dikes outside Shin. A profile through camp is thus (Fig. 3.54).

My ideas of yesterday in regard to H.V. are confirmed by this view. It is in bubble form or tumor form; the strata being

Figure 3.53. Dips on the west flank of Mt. Ellsworth.

Figure 3.55. Continuation of profile shown in Figure 3.54.

Figure 3.54. Profile through Gilbert's camp (flag at base of Shinarump) on the west side of Mt. Ellsworth. "The dip of the Choc. [= *Moenkopi Formation*] is 42° and it shows a tendency to arch toward the mt."

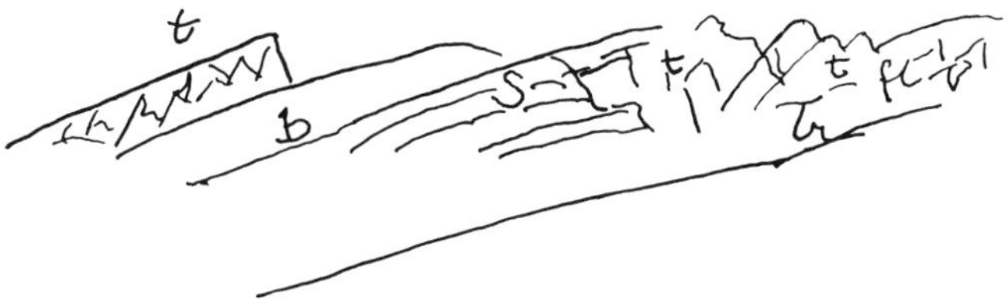

Figure 3.56. Sandstone cut off by trachyte near summit of Mt. Ellsworth.

nearly level on top and the crests controlled by dikes which are radial. Its chief mass is G.M.

Higher I come on another installment of Chocolate and ahead I can see the white Carb. sand [*White Rim Sandstone or Aubrey*] in the cañons. Its dip is less, say 25° or 30°.

At 23.05 = 7960, I am on top of the white sand and find the dip 37°. Camp = 22da (Fig. 3.55).

On the next spur to the W I see trachyte outside not merely the Shin. but several hundred feet of Vermilion—suspect a radial dike.

{p. 55}   At 22.98 = 8170, I am on a second portion of white sand and the dip is 25°–30°.

There are oblique as well as conforming dikes. [*The expression "conforming dike" apparently was Gilbert's term for what today we term "sill."*]

k) [?]   At 22.84 = 8370 I can see across the cañon to the E the bed of sand I last noted terminates against trachyte (Fig. 3.56); and beyond I see no sandstone nor has the trachyte the same tabular character that it has when interleaved. I command a large amphitheater and the negative is strong that this is the end of the sed and that here begins the lava core or chimney [*the stock*]. From *k*, camp = 19da (about) and summit = 23^{ea}. The value of the evidence is weakened by finding slaty sandstone at 22.64 = 8570. Its dip (19°) is not in the line of the section but is to the W. The same rock continues to strike to 27.54 = 8200 where it is chloritic and metamorphic in appearance and is so intermingled with the lava that I fancy it fragmental and detached. [*Shattering*

and metamorphism help distinguish the stocks from the laccoliths.] Summit for this last point = 18^{ea}. The same rock I see again in a small patch still higher and again at the very top.

{p. 56}   Summit of Ellsworth at 12.25 = 22.25 = 9040.

The WSW spur shows seds highup on its side dipping W.

In the SE spur are rocks dipping SE at 12° (1st exposure) and at 16° (2nd exposure). At the base (14da) I see Vaga. dipping about 16° and the G.M. this side prob. has a high angle.

Following the crest to the S where it forks I see that the S Spur has nearly level strata on its crest and halfway down are strata inclining SW at 20°. Through the S Cañon I see G.M. dipping SSW at 25°.

All the seds near the crest 'seem' altered.

From a point near the head of Waterpocket Cañon S to near the Vermilion Island the crest of the white G.M. is a descending line.

As seen from here the D Cliff [*south edge of Tarantula Mesa*] makes a straight line as far as I can trace it and I cannot see its upper surface. C [*Emery Sandstone*] curves up both right and left and I can see its upper surface except near Pennell where its dip is from

{p. 57}   me. Its inclination seems rather from Hillers than Pennell although it is thin on the flank of Pennell. I think I see sed at one place on the SW flank of Pennell capped by lavas and dipping at 40° to the SW. Cliff B is spread so low that I see its

Figure 3.57. "Double synclinal in Cliff B [= *Ferron Sandstone*] as seen from the summit of Mt. Ellsworth" [*W.P. =Waterpocket Fold*].

whole top. It makes a double synclinal about as drawn (Fig. 3.57).

Cliff A [*Dakota Sandstone*] runs entirely clear of Ellsworth but so near its NW base as to be affected in dip. Facing the SE it runs through the pass bet. H.V. and Hillers and then turning more to the N turns the E base of the E spur of Hillers. The Vagabond rocks at the last point make a second step. They moreover fill all the space from Cliff A to Ellsworth and H.V. and cover a broad stretch S of Cliff A and Ellsworth. Beyond the Colorado I do not see them.

On both banks of the Colorado the G.M. is all red but with a difference of degree (Fig. 3.58).

{p. 58}   Returned to the summit.

From NE to E at the base of the Mt. I see G.M. upper and lower 'curving' up toward the Mt. with a max dip of 20° which measures also the depression of the base of Verm.

G.M. extends everywhere from the base of Ells. and H.V. to the Colo. Cañon [*see Figs. 9.12, B; 9.E*] except that a cañon from the divide between the two may cut the Shin. The Colorado does cut the Shin. and the many mesas beyond are of Verm. This remark applies from NE to E. From E to SE and beyond there is much upper G.M. and the tables are larger with less even tops.

The H.V. dip extends to the Colorado [*see Fig. 0.C*] and at its nearest bend a little beyond. On that line (a NE one) the highest rock is Verm. sand.

Toward Abajo the rocks rise—hence from the right and from the left, but Abajo is beyond what seems the culmination of the swell. It is a swell of exceedingly gentle slopes and great horizontal extent. It extends even to the San Rafael Swell. Beside this is a slight dip from a line beyond the D.D.

{p. 59}   river toward the Henry Mts. or W. This is one with the W dip from the Abajo Swell. There is a slight sign along Cataract Cañon judging by the Shinarump.

Just this side the junction of the D.D. and Colo. is a mound of upper G.M. on the Verm. Mesa.

The G.M. in the top of H.V. is almost level or quite [*see Fig. 9.21*]. It is a low angled bubble. Hillers is a high angled (Fig. 3.B and 3.48) and Ellsworth strikes a mean.

I don't understand the NE side of Hillers.

The high plateau N of the D.D. is Vermilion. The Shinarump stands out well from its base toward both the D.D. and Colo. and is as bold a cliff as it is near camp.

The B Cliff shales, without their cap of sandstone, underlie camps 38 and 39 and are the "bed rock" on which runs the water which affords our spring. It is protected by trachyte debris.

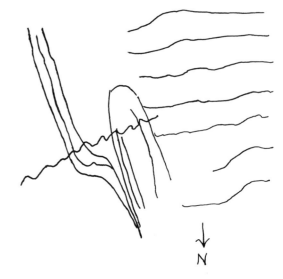

Figure 3.58. "At about its intersection with the Colorado, the Escalante fold passes to the east by echelon, the river cutting the two parts where they are about equal." As viewed from top of Mt. Ellsworth. Notebook 3, p. 57–58.

Ellsworth trachyte is identical with Hillers trachyte.
Tuesday, August 24. Mt. Camp = 25.00 = 5830 = 4.50 am
Belted Cliff dips 2½°.

```
Altitude corresponding to bar. inches--unreduced
```

Inches	Feet	
18	14,000	
		140 for 1/10 inch
19	12,600	
		135
20	11,250	
		130
21	9,950	
		125
22	8,700	
		120
23	7,500	
		115
24	6,350	
		115
25	5,200	
		110
26	4,100	

```
1°  = 1/57  =    90 ft to the mile
2°  = 1/29  =  180
3°  = 1/9   =  280
4°  = 1/14  =  380
5°  = 1/11  =  480
6°  = 1/10
7°  = 1/8
8°  = 1/7
9°  = 1/6
10° = 1/3
```

General note on terraces. Those of the 'munnrei's' field are not built of debris but of rock in situ capped by debris.

CHAPTER 4*

Chapter 4. Notebook 4 begins base of Mt. Ellsworth, Aug. 24, 1875. Aug. 25 to Trachyte Creek passing east end of Sawtooth Ridge and Black Mesa. Aug. 25, sketches of Hillers, Pennell, and Ellen as seen from the butte (Farmers Butte?) near camp by Trachyte Creek. Working up Trachyte Creek and its forks Gilbert records for first time that the lavas are injected. Aug. 27 and 28 in Pennellen Pass and on south flank of Mt. Ellen. Observes west dip of the Cretaceous sandstones. Aug. 28 reaches Ellen Peak. Looking west he notes that Miners Mountain (his Round Top Mesa) is the highest part of the fold (Teasdale anticline) and notes, Aug. 29, that the lavas on Thousand Lake Mountain are younger than the folding. Camp that evening "at a villainous creek tributary to the Dirty Devil. . . .The water in camp is very bad . . . it has run through the Leaden shales (Mancos) and been poisoned. . . ." Aug. 31 inspects Red Arena and then returns to Dirty Devil and finds that route through the Capitol Reef "would be impassable at a high stage of the river. . . .We cross the river 88 times." Sept. 2 back at Supply Camp. Sept. 3, 4, and 5 travel north along west side of Thousand Lake Mountain.

---

*Consult Table 0.1 and appendix for abbreviations.

85

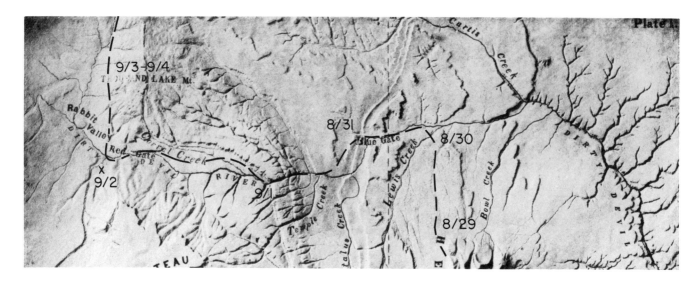

Figure 4.A. Landform maps showing Gilbert's route from Mt. Ellsworth northward through the Henry Mountains (facing page) and westward up the Fremont (Gilbert's Dirty Devil) River en route back to Salt Lake City above. See also Figure 0.D for route in Henry Mountains.

Figure 4.B. Henry Mountains diorite porphyry with inclusion of amphibolite (from U.S. Geol. Survey Prof. Paper 228).

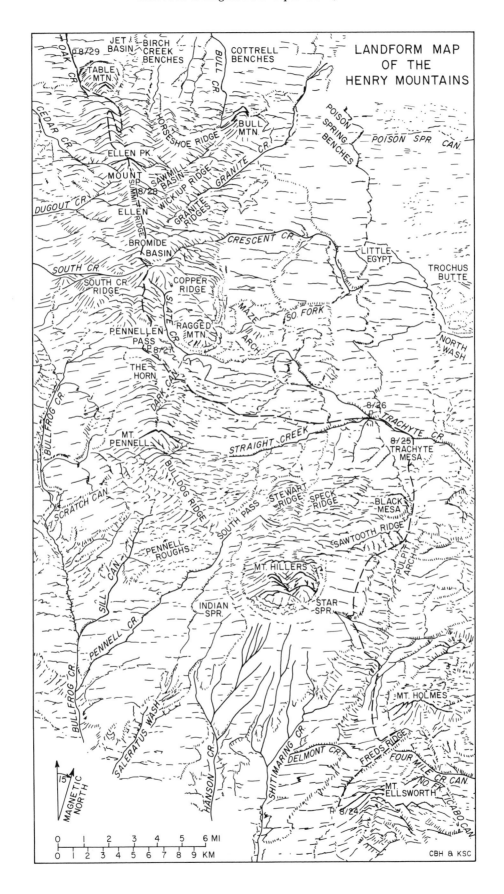

LANDFORM MAP
OF THE
HENRY MOUNTAINS

CBH & KSC

G. K. GILBERT, POWELL SURVEY, BOX 806,
WASHINGTON, D.C.
NOTEBOOK NO. 4
OPENED AT CAMP ON THE BASE OF MT.
   ELLSWORTH, UTAH, AUG. 24, 1875
CLOSED ON THE E BASE OF MT. HILGARD, SEPT. 5,
   1875

{p. 2}   **Tuesday, August 24, 1875.** Camp at base of Mt.
Ellsworth = 25.00 = 5830 = 4.50 am.

In the saddle bet H.V. and Hillers the Belted Cliff dips 2½°
to the NW. Aneroid at base of cliff on saddle = 25.93 = 4370.

Camp 39 = 25.17 = 5670.

The datum point for the identification of rocks in the Bubble
Mts. [*Henry Mts.*] has been *"Gryphaea"* which I have found with
fossils.

I have omitted to mention the frequent occurrence at the
base of the Leaden cliffs of balls of clay rolled on the principle of
snowballs either down the slope or along the arroyos.

A mile before reaching Jerry the section is (Fig. 4.1).

Jerry contains Belted and *"Gryphaea"* rocks and is filled
with *vert.* and *hor.* dikes. The rocks in Jerry

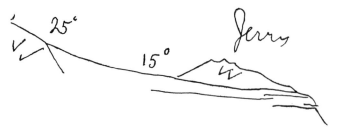

Figure 4.2. "On the pass there is little lava." I am not sure just where
Gilbert drew this section.

Northward my foreground is of strata dipping from
Hillers—strata unidentified.

Beyond Trachyte Creek is a fine study in structure. The
structure lines run NNE. The whole eastern base of Henry First
(H.I.) [*Mt. Ellen*] is an escarpment of strata dipping about 10° to
the W and supporting a great weight of lava, anticlinal with this
are E-dipping strata [*Maze Arch*] with 30° of dip carrying the
whole down into the depths. Then they rise again with a measured dip of 3° (Fig. 4.3).

Figure 4.1. Section a mile south of Jerry.

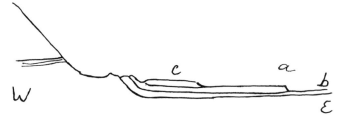

Figure 4.3. Gilbert probably intended this as a generalized cross section
from Mt. Ellen eastward across the southern part of the Maze Arch. He
wrote, "At *a* is the Belted Cliff and the *b* rocks are red (Vag.?) (G.M.?). *c*
is gray shale preserved in the syn. of some depth with reference to the
drainage plane. The dip at *a-b* is probably one with the great northward
swell noted yesterday." Notebook 4, p. 4–5.

{p. 3}   dip 10°–15° and with the dip run into Hillers. I see no
rocks on edge in the Hillers adjacent.

Jerry is ENE from Hillers but the dip is ESE. On the pass
there is little lava (Fig. 4.2). N of the spur the dip is more
northerly. It is simply a place in which the flexure of the flanks is
not concentrated. Beyond the pass we descend into red rocks that
look like Vag. They are somewhat interleaved by trachyte, and
lava is visible beyond the foothills and near where the dip
changes. Trachyte Creek should follow a synclinal. The lava on
the plain [*Trachyte Mesa*] lies flat and I take it to be injected. It
may however be ejected.

On Bell's trail station No. 6.

It is a crest of the foot slope—trachyte over sandstone. The
sandstone is dipping away from Hillers at a low angle.

This eastern aspect of Hillers is altogether different from the
opposite. A thousand feet above me (Sta. 6 = 25.10 = 5750) are

{p. 4}   level strata of white sandstone. They make a shoulder
on the spur nearest and are floored and roofed by lava. The lava
below is in horizontal leaves and to the right and left (Jerry etc)
curves downward on the slopes.

{p. 5}   I can trace the structure from Trachyte Creek for about
10 m. N. Then it must turn or die out. At the last point seen it
looks as though it turned and at the same turn diminishes in the
eastward dip.

The trachyte crags above the seds look massive and show no
horizontal (or other) structure.

Pennell I make nothing of except that its eastern base has
strata dipping rather toward than from it.

Camp 40 (25.13 = 5210) in Trachyte Creek (branch from
Hillers).

Wednesday, August 25. Camp 40—25.27 = 5560 = 7 am.

Station 1 is 1/3 mile NNE from Camp 40. No rock in place.
25.27 = 5560. [*The morning barometric pressure is suspect; it is
identical to Sta. 1 and would indicate a very considerable over-
night change.*]

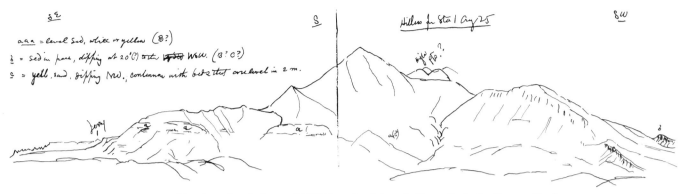

Figure 4.4. "Hillers from Sta. 1, August 25." This station 1 is probably on the butte, Farmers Knob, a half mile SW of the site of Trachyte Ranch. The level sand *aaa* noted by Gilbert is one of those forming the upper part of the Morrison Formation. *b* is probably the Emery Sandstone in South Pass, the pass between Mts. Hillers and Pennell. Its dip is N of W.

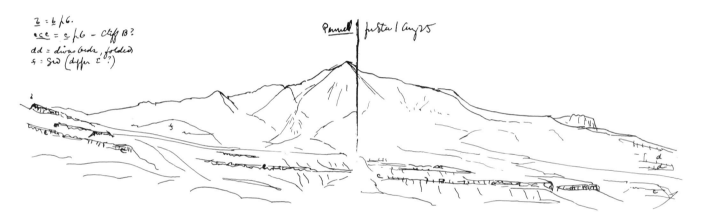

Figure 4.5. "Pennell from Sta. 1, August 25." Compare Fig. 4.4. Three ledges of sandstone to be seen from top of Farmers Knob are the Dakota (Gilbert's *"Gryphaea"* Ss), the Ferron Sandstone (Gilbert's Lower Sand), and the Emery Sandstone (Gilbert's Gate Ss). South of Pennell Peak is the canyon of Straight Creek, which is incised into the south edge of the Mt. Pennell stock. The top of the Horn laccolith shows on the skyline middle right.

I think the beds under Belted Sta. 6 of yesterday were Belted, and for the present call the bed above in the face of Hillers (as p. 6) Cliff B contiuous with *c* (p. 6) and curving monoclinally at *a*.

{p. 8}   I think the Henry fold [*Maze Arch*] runs into Pennell at *d-d* (p. 7) (Fig. 4.5).

I cannot make out the *"Gryphaea"* sand although I see the purple shales it should cap. It must be inconspicuous. The Belted Cliff E of the syn. is unmistakable. It runs 1 mi. E of Sta. 1 [*1875*] and can be seen 8 mi. N holding a very straight course. From 4 mi. N to 7 mi. N it is bared of all superior beds clear across the syn. but beyond and in the foreground it is covered by gray shale [*Mancos*] the upper surface of which is a drainage plane sloping from Henry I [*Mt. Ellen*]. E of the Belted [*Morrison Formation*], the Vagabond beds [*Entrada Sandstone*] make a bench of their own. It is very uneven in outliers and is chiefly preserved like the

Leaden clay [*Mancos*] by an old limit of drainage. G.M. beyond extends to the cliff of the D.D. and is resumed beyond it.

A section N and S or parallel to the Henry Mt. general trend, and along the E base will not show a displacement of note. Whatever its undulations they would be slight and so far as I know now it will be safe to make that a base plane for the Henry displacements.

At 25.60 = 5200 I have *"Gryphaea."*

At 25.85 = 4960 I am on top of the Belted Series having

{p. 10}   passed 2/3 blue gray shale and 1/3 painted shales. The *"Gryphaea"* horizon is either lifted by the introduction of gray shale or the species is higher than I have found it this season and its familiar sandstone absent. At this point I am in a branch of Trachyte Creek rising on Hillers and Pennell. (Note I did not read on the highest of the Belt conglomerates but 30 ft. lower.)

Figure 4.6. "Henry I from Sta. 1, August 25 [*1875*]. See p. 13 [*i = Bull Mountain, Gilbert's NE Butte; X1 = Ellen Peak*]."

Across the 1st creek I climb a knob capped by quartzite, a local unexplained occurrence. It is in place resting on gray shale.

(The mingling of sed. and lava that I noted in approaching Pennell the 1st time, in the light of later facts, would seem to be an interleaving of dikes [*and sills*].)

Sta. 2 (25.77 = 5000) [*1875, not 1876*] is on the S bluff of Trachyte Creek where it receives a tributary from Pennell.

I can now make out more plainly the *"Gryphaea"* horizon. It is marked by a sandy streak that occasionally develops as sandstone. The shale below it is not the legitimate blue gray of the Leaden series but is pea-green and shows traces of purple all

{p. 11}   the way down to the full purple—100 ft below the sand. Above the sand is the leaden gray weathering yellow.

The Hillers-Pennell [*Pass*] how looks as sketched on p. 12 (Fig. 4.7).

I can trace the Henry fold quite to the middle of Pennell as seen from here.

Where I strike Trachyte Wash it has no water (25.87 = 4930) but we soon find it rising in spots on the gray shale. The highest rock the creek runs on is the gray shale near its base. I find *"I. problematicus"* and *"Ostrea"* near the base of the B Cliff series [*Tununk Shale*] (25.71 = 5090 = 11.30).

Loc. 918 Section of fold
1 = gray shale	100 ft+
2 = yellow sandstone and conglomerate with *"Gryphaea"*	50 ft
No. 2 dips 12°	
3 = gray green and purple shale	150 ft
4 = white conglom. and sand—the Belt series (with red shale)	
Lunch at top of 4 = 25.60 = 5210	
5 = red and gray aren. shales	
6 = dark and pale red sand (Vag.)	
7 = dark red zone = Jura	
8 = white G.M.	

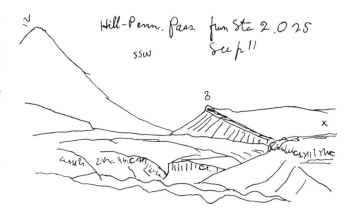

Figure 4.7. "Hill-Penn. Pass from Sta. 2, SW. See p. 11." *b* is Cliff C, = Emery Sandstone, dipping WNW at 10° or 12°, *ccc* is therefore fixed as Cliff B, = Ferron Sandstone. The indicated direction seems too far for the south as if Gilbert was referring to magnetic north. See introductory chapter for discussion of declination.

From *b* (see Jura) I can extend the identification of beds through the Henry fold but cannot make out the fold well. It is

{p. 13}   headed here by a syn. that runs from the Divide down Trachyte Cr. but resumes with gentler slopes under Pennell which Mt it runs directly into. Northward too the line of max flexure makes an excursion westward and returns to its eastward position—about 3 miles away this occurs. The most distant point I see is due N.

In the sketch (Fig. 4.6) on p. 9 is upper G.M.

hhhh is Belt Series and the greater part of the remainder is Vagabond.

C and *a'* = 25.40 = 5430

Crossing over the divide fork of Trachyte Creek and climb-

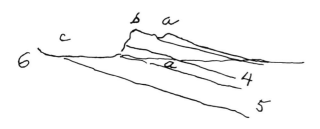

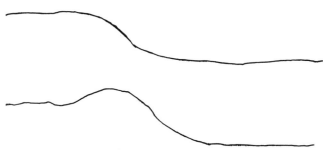

Figure 4.8. Cross section on p. 15. At *a* dip = 22°. *a* = 25.20 = 5630. *c* from *a* = 16.5^{da}.

Figure 4.9. Cross section of the Henry fold (Maze Arch) referred to on p. 15.

ing the slope of Pennell I finally find water and encamp with Mr. Adams on a branch of the wet tributary we crossed this morning. **Thursday, August 26.** West Camp—24.49 = 6420.

Climbing the hill to the S I am not far from Hillers and see no seds in this face of it. The hill I am on is strewn with immense blocks of yellow sandstone—not in place but probably not far out. The hill is at *x* in the sketch of page 12 (Fig. 4.7) and the sandstone very likely belongs to Cliff D [*Emery Sandstone, not Mesaverde*].

The slope of Pennell

{p. 14}   I rode up yesterday is of Belted (in place) and perhaps *"Gryphaea."* Its dip swings from E at the creek to S at the crest of the fold at which latter place it is about 3° S of it to Wet Camp. There are few and small exposures but the Pennell debris must be underlain by Cret. all along its eastern base.

From all I have seen so far it would appear that the Henry fold is independent of Henry I. The rocks W of the fold run into the mountain base with a dip toward the Mt. and have a great descent to make in descending to Pleasant Creek [seems to refer to *Crescent Cr.* on the benches north of the maze]. The meandering of the fold (= its intersection by synclinals) is a surprise, it looked so straight at first sight.

The crest of Belted I stood on yesterday (b) and especially the northward prolongation of the same are levelled (or as by) an old denudation limit [*pediment gravel*]—sloping from Henry I.

The east facing Belted Cliff that, seen from the end looked so remarkably straight, from this point [*seems*] to be cut by mountain drainage into a series of islands.

Later it appears that hill or ridge I climbed is composed of beds dipping 8° or 10° and directly away from Pennell. The only exposures in place are lava but they are evidently injected beds.

{p. 15}   Rising toward the pass and striking the trail of the party I find I have crossed Cliff B [*Ferron Sandstone*] and am on an injected lava bed (conformable). [*This is Gilbert's first note indicating that the igneous rocks in the Henry Mountains are intrusive and not lavas. Evidently an old trail extended around the northeast side of Mt. Pennell and Gilbert saw the roof of the Coyote Creek laccolith or possibly the roof and floor of one of the*

sills above it.] C Cliff is not carried up the mountain slope as a cliff but its beds follow up to at least this height (23.65 = 7370), keeping 1 to 1½ miles S of the outcrop of Belted. I suspect that a sandstone just below this lava belongs to the C series [*Emery Sandstone*].

The Henry fold [*Maze Arch*] is too complicated for a sketch from this direction. The dip with which the beds start into the Mt. is very gentle. I see from here nothing more than 2°. At the G.M. Head [*The Block*] however, there is locally an increase that makes the mono. and anti. The diagram exaggerated is as in Figure 4.9.

Beyond the farthest pt. I saw yesterday (N) the fold seems to turn westward but I cannot trace it far.

Camp 41 = 23.12 = 7990.

I have now seen the lava of all the 5 Henries and find them essentially identical—The same white

{p. 16}   feldspar and the same black hornblende (Fig. 4.B) and nothing else either in paste or contents. Unaltered the rock is always gray and usually pale. Oxidized it acquires flesh tints and its soil is pale amber. In decomposition the paste usually survives the crystals. [*See Editor's note, Gilbert, p.17 of Notebook 9.*]

I think there has been a little local faulting in the vicinity of the G.M. Head [*The Block*] E of Henry I—faulting that has lifted the Head a few hundred feet.

In the notch of the fold beyond the Head, the Belted escarpment (dipping) far to the W. In the notch this side of the Head the Belted hogback holds its course undisturbed and only the lower beds (Vag. and G.M.) show the deflection of the crest of the fold.

Belted rocks under the ragged butte [*Ragged Mountain*] across the creek cañon must run nearly level (Fig. 7.21). The top of the Belted is there (say) 300 ft lower than camp 41. I see no seds under H.I. that seem higher than camp 41. [*Gently dipping sedimentary rocks, poorly exposed, that are higher than Gilbert's camp are extensive around the southeast and south slopes of the shatter zone around the Mt. Ellen stock.*] There are some nearer the pass and against

{p. 17}   H.I. that I think higher geol. than the Belted. The

Belted seem to hold the same level and dips on both sides Trachyte Cañon near the pass.

Friday, August 27. Camp 41 = 23.05 = 8070

This morning Mr. Bell is detached (with Sorenson, Adams and 2 packs) to rejoin at Supply Camp on September 2 or no later than noon September 3. We march this morning and he remains in camp to give Pennell another climb tomorrow.

At 23.45 = 7600 as we approach the pass we are on a level with a sandstone opposite that can only be one of the Leaden cliffs. It dips 4° to the SW (or toward Pleasant Creek) [= *Crescent Creek? see p. 4–14*] and also toward Pennell. It appears also on this side but in slides so that I do not know its dip. Bet. this outcrop and craggy butte [*Ragged Mountain*] are exposures of upper Belted and *"Gryphaea"* dipping at a greater angle toward

{p. 18}   Pennell. The Belted under Craggy Butte is 200–300 feet lower than I am now and its dip is 2° to 3° to the W. [*Without saying so, Gilbert is keeping track of the Belted, the Morrison Formation, in case he finds it becoming exposed again at the next hill.*]

The bald face spur of Pennell that faces the pass [*The Horn*] and is close at hand shows horizontality on top as though it might be a conformable dike. [*Sheeting at the top of the Horn laccolith, his "conformable dike," represents further recognition that the igneous rocks are intrusive.*]

Yesterday I saw slate mingled with lava debris as high as I climbed—500 ft above camp 41.

Later I can trace on the Pennell side of the pass at a little higher level an easy westerly dip but cannot see the materials of the shoulders that indicate this.

Starting up H.I. [*Mt. Ellen*].

The two dips unite or nearly unite on the pass. The rock is Cliff B [= *Ferron Sandstone*]. It dips to the west all the way 4° to 6° and also from each mountain to the pass.

Cliff C [*Emery Sandstone*] is deeply scored by drainage from the pass. It dips from the pass, from Pennell, and from H.I. It runs well up

{p. 19}   on the flank of Pennell W. of the summit and I measure 7° dip with evidence of a slight increase mountainward. To the W from here it is 2 m. distant and with a dip of (10°–15°) westward.

The D Cliff [*Mesaverde Formation capping Tarantula Mesa*] beyond barely begins to dip in excess of its plateau slope. Where it is cut off, both crests are lower than the pass in this latitude.

Ahead I see a bench structure on H.I. that may mean the presence of C [*Cliff*]. If so it dips 5°W and 5°S in strict sympathy with B in the pass.

Pass = 23.68

There is sandstone and gray shale at the shoulder (mentioned above) on H.I.

The hogback of C runs along the base of H.I. 3 m. from the pass and then is lost under debris. It reflects a max of dip and the

section from where I sit through it would be as in Figure 4.10. The max dip is about 10°–12°

Figure 4.10. Cross section of Cretaceous formations dipping WSW off the south part of Mt. Ellen, as seen by Gilbert from Pennellen Pass. If the indicated trachyte is the South Creek Ridge laccolith, it should be shown on top of the Ferron Sandstone (B).

{p. 20}   and runs quite outside the trachyte. So the Henry I is between folds and not of them.

The ground up to this point has much gray shale—hardened (22.06 = 9260).

A little higher I find gray shale unaltered and in place and it seems to form the great mass of the first Mt. mass about which we skirt. Its dip is westerly—3°, 10°, 15° observed). The ground all along the slope has more shale than lava and the latter in this part is only in dikes [*shatter zone south of the Mt. Ellen stock*].

The spur by which we climb to the summit is determined by a dike, a dip noted on it low down would rise to the summit [*Gilbert climbed the south rim of the Bromide Basin, location of the Mt. Ellen stock, and shatter zone south of it*].

The cañon westward from the crest we follow reveals in section at the Mt. base an anti. arch. [*Could be any of the several laccoliths near the mouth of Dugout Creek.*]

The E Henry fold curves NW so as to just touch the base of butte *i* (Fig. 4.6) (p. 9). At that point the E facing cliff of Belted approaches very near. The fold as marked by the Belted maps as on the next page (Fig. 4.11).

Summit = 20.95 = 10,670 = 1 pm.

March 8 am to 1.50 pm. Saddle near camp 21.60 = 9880.

Camp 42 = 21.92 = 9460. The water on which we encamp has no willows nor aspen about it—only firs.

There is no valuable timber on the Henry Mts. Large portions of their surface are treeless, and the only considerable groves are fir and aspen. *Pinus ponderosa* forms a few small groves and on H.I. I note a pine of shorter leaf and ranging with the high firs (10,000–11,000 ft). It is the same we found on Oyster Cr. and grows deformed for the most part, yielding to westerly winds. I see one stunted trunk 3 feet in diam.

{p. 22}   The slopes are so steep about our camp that the sun sets at 4.15. The mountain top is covered by trails—mt. sheep? One zigzags very nearly down a hill to this spring and I measure its grade (28°) assuming that a trail winding down a smooth hill

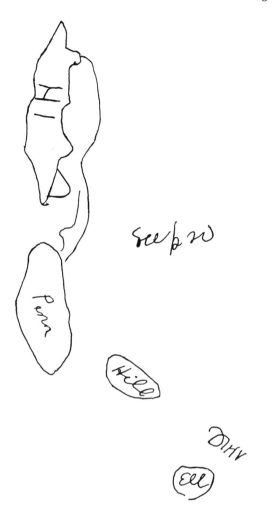

Figure 4.11. Map on p. 21 shows relation of the fold east of Mt. Ellen = H-1.

will give the max grade convenient to the animal. Man would seek an easier grade.

Near the saddle we passed another exposure of shale showing dip—still westerly.

Saturday, August 28. Camp 42 = 21.95 = 9410

pt. × 21.97 = 10,650

Bar 21.44 on the N Point of Henry Mts. Graves "point x" of sta. 27—24 ft below summit.

Hours	AL		DT		2144	Aner 3	
9 am	60		57		19.858	20.99	
10	61		62		19.867	20.976	radiation from rocks
11	66		64		19.884	20.975	good
12	62	1/2	59		19.863		
1 pm	62		58	1/2	19.866		
2	62		59	1/2	19.866		
3	59		59		19.856		
4	54	1/2	56		19.860		
Camp 42 at 6:15 pm = 21.89 = 9490							

{p. 23}   The high slopes of the mountain are so smooth and rounded that dips are not revealed in the distance and color cannot be depended on to distinguish trachyte and sed masses. In following the creek we travelled as much on slate as lava and once struck sectile sandstone. On the point just south of this and in the intervening saddle are slates (plainly Cret. shales altered) dipping 60° to the N.

The C Cliff [*Emery Sandstone*], except where obliterated by debris slopes, is a hogback all along the W front standing entirely clear of the lava in place, = from one to two miles separated indeed. Cliff B [*Ferron Sandstone*] shows little but its tendency is the same. It shows close to the trachyte and I suspect it makes toward it like the Belted at the other base of the range.

At this base of 'b' (p. 9) (Fig. 4.6) I can see an anticlinal structure of *"Gryphaea"* sand—looking as though the fold at the E ran straight through that butte [*Bull Mountain*] on its way to circle the N end of the Mt. [*Ellen*]. There is little visible at the north to indicate structure but I an confident that where the B Cliff (E facing) comes from the north it rises to the mountain flank. The Henry trachyte is contrasted with the Aquarius by its weathering as well as color. Its angular blocks hold their edges and round boulders

{p. 24}   are not developed except by much rolling. Its disintegration proceeds by subdivision rather than by the reduction of its blocks.

The steepest slope I see in profile measures 30°. The E slope of this point measures 33°–34[ada].

I cannot identify Cliff C N of Needle Butte [*Factory Butte*].

Cliff B starting from the N end of the Mt. and facing E runs NNW across the D.D. River toward San Rafael Swell then circles broadly about Needle Butte and returns along the W base of the Twin Mesas [= *North and South Caineville Mesa*].

Cliff A [*Belted*] running a mile or two E of the NE butte of Henry I [Fig. 6.D] continues with diminished strength and height to the north (E facing) where it is much rent by cross drainage. Meeting the S.R. [*San Rafael*] Swell it turns back SW along its base and after following that to its S end runs with northward face toward Mt. Hilgard. It may disappear for a time under the trachyte or it may turn N and resume its relation to the S.R. Swell (between its western excursion and the S.R.S. there are a half dozen dark topped buttes of miscellaneous dip that I am at a loss how to refer). [*These are the analcite diabase dikes and sills described by Gilluly, 1927, 1929. See Reference in Introduction.*] There are two other divisions of the Belted Cliff in sight. One {p. 25}   (the type) after it is cut by Pleasant Creek [*Crescent Creek, north to San Rafael Swell?*] [*Another*] follows the base of the Escalante fold to M.L. or rather to the ridge N of M.L. [*Thousand Lake Mtn.*] upon the slope of which it rises. Perhaps it is so cut through as to join the N facing division at that place. The third division is an inward facing circle lying W of the S Twin Mesa (the Red Arena) a small quaquaversal of which I shall see more. At present it seems to occupy a triangular area between three synclinals—one, that which contained the Twin Mesas, and

Figure 4.12. "San Rafael Swell from Sta. 27." Gilbert's station 27 is on top of the North Caineville Mesa. D.D.R., Dirty Devil River, just east (right) of the Caineville Mesas (C); B = Ferron Sandstone extending around Factory Butte = Gilbert's Needle Mountain.

the others separating it from the Escalante fold and S.R. Swell, respectively. The two latter join westward to run toward the Blade or Hilgard.

The Escalante fold has its maximum in the Roundtop Mesa [*Miners Mountain*]. Thence it curves to the left to reach M.L. The synclinal Howell noted in its crest has its greatest intensity at the D.D. River or just N of it and fades both ways. Toward M.L. it is parallel with the fold but in the opposite direction it runs out of it near Temple Creek—runs out and at the same time disappears. Both lines of G.M. cliffs rise toward M.L. and cover more of the fold. At the Roundtop Mesa the Carb. is the crest of Escalante and D.D.

{p. 26}   folds at once. N to the D.D. the Escalante is crested by Carb. and thin overlying Chocolate. Beyond the D.D. the crest is doubled by the syn: the west crest bearing successively Chocolate, Shin., and Vermilion with a little Gray Cliff under the protection of the trachyte and the eastern beginning with the Gray Cliff and holding it throughout.

The top of Aquarius is a wonderfully level line from this point of view.

The M.L. lava is subsequent to the Escalante fold, 'unquestionably'. [*The folding is older than the lavas; the Thousand Lake fault is younger. See discussion of regional structure in introductory chapter.*]

The S.R. Swell is spread before me in a beautiful and comprehensive way. I can see almost the whole of its oval disc. The gray Cliff of the G.M. completely encloses it. I can see the Verm. Cliff on the further side and it is probable that Carb. strata are laid bare by the uplift and erosion. In a general way the magnitude of the uplift is measured by the distance from the top of the Carb. to the top of the lower or middle Leaden Cliff (B or C). The sketch I have attempted on the next page (Fig. 4.12) is too '   ' for the page that I have to 'penetrate' and too much foreshortened for the altitude I have, but is not false in its general expression.

The westward dip from the Abajo Swell [*Monument Upwarp*] extends to the S.R.S., extends to the Henry Mts., and bet. the S.R.S. and H. Mts. extends to the Twin Mesas. It is everywhere gentle.

I am impressed today with the idea that the swells, the anticlinals,

{p. 28}   are in this region the entities and the synclinals are the intervals (inter'vales'), or residuals or corollaries—non-essential though unavoidable adjuncts.

In the Red Arena are exposed a floor of the Vag. and an amphitheater of Belted Cliff.

Sunday, August 29. Camp 42 = 21.88 = 9500. Marches 6.50–11; 2–8.30.

Climbing the divide and descending to the west I follow by frequent outcrops of sandstone and comparing dikes a westerly dip 10° and 15° at first and diminishing to 4° at 23.10 (= 8,000). This continues for a mile and then I note a local reversed dip. The shales now are unaltered although traversed still by dikes and covered (N) by a heavy lava mass.

On the next page is a diagram (Fig. 4.13) of the section on the N side of the cañon. The direction of the cañon is nearly W. Besides the dips shown there is at this point a northerly dip of 10° (about) and this N dip soon disappears toward the north. *"Gryphaea"* and B. are at the base of the Mt. as hogbacks. C is locally covered but appears as a hogback farther to the N and to the S.

On the S side of the cañon appear 200 ft of Vag. sand. The Belted is not thick—at most 200 ft.

Looking S from the mouth of the cañon I have the following section—(Fig. 4.14).

{p. 30}   I think the section N of the cañon is substantially the same but at the cañon the whole thing (east of x) is thrown down sharply (or left behind) thus—with a compound fracture (Fig. 4.15).

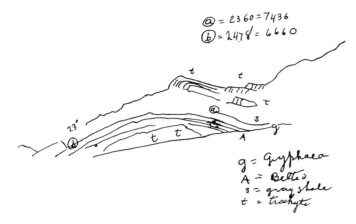

Figure 4.13. Exposures of sills in the Red Arena area. Notebook 4, p. 29.

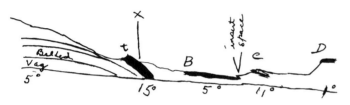

Figure 4.14. Exposures of sills in the Red Arena area. Notebook 4, p. 30.

Figure 4.15. Exposures of sills in the Red Arena area. Notebook 4, p. 30.

In the next cañon N and the next the Vag. rocks are exposed and the principal rock of the hills intervening is Belted. I think the general structure is monoclinal as in sketch (Figs. 4.14 and 4.15) but it is so disturbed by lenticular lava masses that it is hard to say what is normal.

{p. 30}   At the N end of the Mt. there is a decided tendency of dips to circle around but much is covered by debris.

Badlands of a typical characteristic come in our way below the C Cliff as we approach the D.D. They are not too soft for travel.

Cannot see the structure of N end of Henry I. The N spur is strangely buttressed with little turrets all about the base as though it might be a little Hillers but no seds show there. [*Gilbert is looking at the faulted north and west sides of the Table Mountain bysmalith.*]

Camp 43.   26.26 = 4520 is on a villainous creek [*lower end of Oak Creek*] tributary to D.D. river just below Twin Mesas. It is at the B sandstone [*Ferron*] and there are 5 thin coal seams in sight. The badlands we crossed abounded all through the C shale [*Blue Gate Shale*] in *"Inoceramus"* fragments. Near lunch camp there is a thicker (3–4 ft) seam of coal.

Monday, August 30. Camp 43 = 26.35 = 4430. March 7.50–

The water in camp is very bad although good when it left the mountains. It has run through the Leaden shales and been poisoned. [*Magnesium sulfate water that causes, in local parlance, the thin dirties.*]

In the valley of the D.D. between the Twins is a mesa = old floodplain capped with debris from the west, chiefly Aquarius trachyte. It is 200 ft high and there is another below it of half the height. We have just passed the limits between the two trachytes.

We strike the river at 26.28 = 4400.

Camp 44 on the Dirty Devil = 26.30 = 4480.

We are in camp about 1/? of the C. shale. Climbing we have gray shale with *"Inoceramus"* and interrupted fillets of concretionary limestone to 25.42 = 5400. Then by alternation we pass to massive yellow sandstone. Top of mesa = 25.25 = 5590.

Sta. 28 (25.05 = 5800) is on the W edge of the mesa and is the highest (?) point owing to the curling up of that edge in sympathy with the E branch

{p. 33}   of the Escalante fold.

Egglofstein [*?*] took his cañon topography from such a piece of drainage as lies below us in the clay flat with its ramifying arroyos.

An old drainage level has left its terraces and color marks all through the monoclinal valley. I trace it as far as Pleasant Cr. and beyond—and as far west as the Escalante fold. Yes, its branches are to be seen S to the Waterpocket Cañon divide. They ascend toward that point, toward the mouth of Pleasant Cr. cañon, and toward D.D. cañon.

On p. 34 I have sketched some points (Fig. 4.16) that I shall use tomorrow in running a section from the C Cliff in the S Twin to the upper Vagabond rocks in the Red Arena.

The Twins rise more gently than I can measure eastward toward the Abajo Swell [*Monument Upwarp*] and more decided toward the E branch of the Escalante making a synclinal close to the W edge.

There is a perceptible settling of peninsular salients by undermining.

The synclinal fold which

{p. 35}   Howell noted on the Escalante does appear beyond the limits of that fold; it is the mono which makes the E limit of the 'Red Arena'. To the S. of that mono the 'E. branch' has a less throw than to the north for the Red Arena mono does not cross the E. branch. If this is a true correlation of the Red Arena mono and the super Escalante syn. then the Escalante mono should be thrown at the intersection, and its hogbacks deflected (in the N

Figure 4.16. Red Arena area. "Sketch to indicate points Graves will sight on to aid tomorrow's section."

course to the east; in the S to the W). The course of the *E branch* fold is 55°E (20 m) and NNE (4 m).

The greatest dip I can see N of me on this mesa is 10° (to the E).

Needle Butte [= *Factory Butte*] is just in the synclinal bet. Abajo and San R. Swells.

San R. Swell is straight or even concave on the ESE side and monoclinal near Needle Butte with a dip of 30°. The E branch joins it very much as it joins Escalante—with no elbow. Taking the hogback

{p. 36}   of cliff B as an index its course is shown by the diagram (Fig. 4.17). Its sinuosities will appear by the topography.

I still am at a loss to understand the dark buttes toward Hilgard and Musinia. Their substance seems to be Vag. Their dark caps may be remnants of trachyte *debris.* Their dips indicate that the folds bet. Escalante and San Rafael swells are not yet unravelled. It looks as though the Red arena Swell (see diagram p. 35 = Fig. 4.17) might have a duplicate in the Butte Swell (BS p. 35), a duplicate evidenced very differently by reason of the peculiar selection exercised by lava debris.

The Gilson Creek Cañon [*Muddy River?*] cuts across the SRS as the DD cuts the Escalante. It is surely inconsequent. S of it the arch of Jura is complete but N of it the Shinarump is cut away from the crest. [*The Muddy River, and the San Rafael River next north of it, are indeed inconsequent. Both are anciently (Oligocene?) superimposed across the San Rafael Swell. Since that superposition, erosion has carved the huge strike valley at Castle Valley, 20 miles wide and 75 miles long, between the west flank of the Swell and the Wasatch Plateau. Castle Valley is 2,000 feet lower than the surface at which superposition occurred.*]

The San Rafael Swell and the Escalante fold are each worthy of monographs.

There is no trachyte drift on this mesa.

The Red Arena now shows considerable elongation N and S—especially N. Its structural

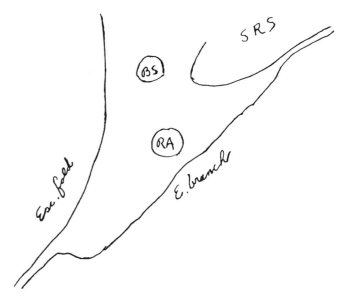

Figure 4.17. Relation of hogbacks seen by Gilbert from the west edge of the North Caineville Mesa, his North Twin. Esc. = Escalante = Water-pocket Fold at Capitol Reef; E. branch (of the fold); RA = Red Arena; BS = Butte Swell; SRS = San Rafael Swell.

{p. 37}   form is as in Figure 4.18, the E. branch making the eastern limit. The southern portion is drained by a special set of washes but the northern is cut through by a wash from the Buttes region and its symmetry of wall destroyed. The diagram shows what might be the outcrop of the upper Belted. The lower Vaga-bond in the southern part surrounds the Arena more.

Bearings to My D [?] points
RP Pennell = 309°11′
  I   337°17′ = dip = 1°
  II   357°43′   error 4′
  III   14°14′   corrected = 56′
  IV   28°55′ ∴ 237 ft.

Figure 4.18. Elongation of the Red Arena.

Graves point on plane table of salient bet D.D. and creek from Henry Mt.—a spur of cliff B bears 148°16′. Depression = 1°44′ (corrected 1°40′).

The distance by plane table from I to Station 28 is 2 2/3 mi.

5280
5280
1760
1760

2 2/3 = 14080 ft

{p. 38}   **Tuesday, August 31.** Camp 44   26.44 = 4330
(Figs. 4.19 and 4:20)

{p. 40}   The general section across the Arena is about as sketched (Fig. 4.20). The surface of the Arena inclines eastward at 3°. The change of dip is near the W margin at (t). The west dip at *S* is 5°. [*The following description only partly agrees with caption of Fig. 4-19.*]

Trachyte from Aquarius is sparsely strewn over the surface and caps all the buttes along the W edge. They are all table topped, the top surface belonging to an older drainage level and inclining E.

Sitting near the S end of the Arena I see A10 in a perpendicular cliff on every side but this—every visible side that is—for I know it is cut through in places.

The B hogback [*Ferron Sandstone*] along the base of the N Twin runs SW and is straight nearly as far as Sta. 28. Then it turns to the S.

I am near the base of A7 and read 25.98 = 4820.

The divide by which I leave the Arena dips 'out' only to A6

{p. 41}   and on that I descend into the synclinal at the base of the Escalante fold. Where I cross the syn. drainage I am in the middle of A3; but to the north of me the syn. contains an oval blue basin of B2 and between me and the D.D. River is a trachyte capped mesa holding nearly all—at least 300 ft—of B2.

Farther north the syn. rises and is made entirely of A rocks—upper and lower. It passes bet. the E fold and the Butte anticlinal.

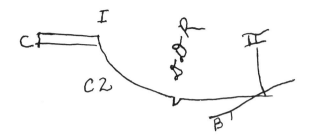

Figure 4.19. Gilbert's section from the South Twin (South Caineville Mesa) to the hogback of Ferron Sandstone and Morrison Formation west of Caineville. (Back of p. 38, p. 39).

C1 = yellow sandstone, massive, 150 ft
No dip (=change of dip) at I
C2 = blue gray shale. Near its base *"Ammonites Baculites, Inoceramus"* (Loc. 920), 1200 ft.
B1 = Sandstone yellow, bedded, with gray and green shale and thin coal. 150 ft.

II
From II, I = 110^{aa}; Sta 28 = 10^{aa}; strike = 353^{aa}; dip = 16°.
B1 resumed, 100 ft.
B2 = blue gray shale like C2, 500 ft.
B3 = green gray shale with *"Gryphaea,"* 10 ft. Dip - 20°.
A1 = slaking conglomerate, 10 ft.
A2 = firm conglomerate, yellow, 20 ft.
A3 = slaking sandstone or shale; weathering in badlands, white above and purple and white below; 200 ft.

III
From III, I = 122^{aa}; II = 118^{aa}; Sta 28 = 28^{aa}; strike = 345^{aa};
dip = 20°.
A4 = yellow conglomerate, 20 ft.
A5 = Like A3, white lead, red, yellow, interrupted above point by a sandstone, 300 ft.
A6 = white conglomerate, 100 ft.
A7 = green and red shale and white gypsum and banded; 300 ft.
A8 = green shale, aren. and 'goodies'; 50 ft.

IV = 23.36 – 4420
From IV, Sta 28 + 44^{aa}; I = [*blank*]; strike = 160^{aa}; dip = 10°.
A9 = white sandstone, 25 ft.
A10 = red sandstone soft as shale. After 150 ft of this the dip is reduced to 7°.
The dip continues to diminish and the deepest erosion in the Arena cuts only about 300 ft into A10.

Figure 4.20. Gilbert's cross section of the Red Arena. West is to the right.

Figure 4.C. View east to escarpment of the Escalante fold (= Capitol Reef) where the Fremont River enters it at Fruita. The gravel terrace along the river is about 500 feet higher than the river and roughly 1,000 feet lower than the sandstone knobs on the fold. Foreground is Moenkopi Formation (= Gilbert's Shinarump Group); the thick sandstones are the Wingate, Kayenta, and Navajo Sandstones (= Gilbert's Great Massive). The prehistoric Indian trail that led through the Escalante fold did not go down the Fremont River because of washout by flooding. It went along the foot of the escarpment and down Capitol Gorge, a truly spectacular drive now closed by the National Park Service.

In the syn. I read 26.09 = 4710.

Climbing the mesa pt looking down on the river I read 25.65 = 5150 and am upon or near the horizon of the *"Gryphaea."* A3, A4, and A5 extend to the river below and a little more, allowing for dip.

The dip of A6 (down below) is not over 6° or 7° and that bed is not over 200 ft thick. The trend is SE by E. There is an ascending dip all the way to the G.M. and at the top of that bed it is much

{p. 42}   greater. The foss. Jura is well represented and caps the G.M. on all maxima.

Descending I find a heavy conglomerate in A5.

Figure 4.21. Measuring A10 [*the lowest unit included in the section, Fig. 4.19*]. x = 25.85 = 4950. Dip at x = 7°. y from x = 1 2/3da. Dip at x = 12°. x′ = y = 26.03 = 4770 = camp 45. Dip at y = [blank].

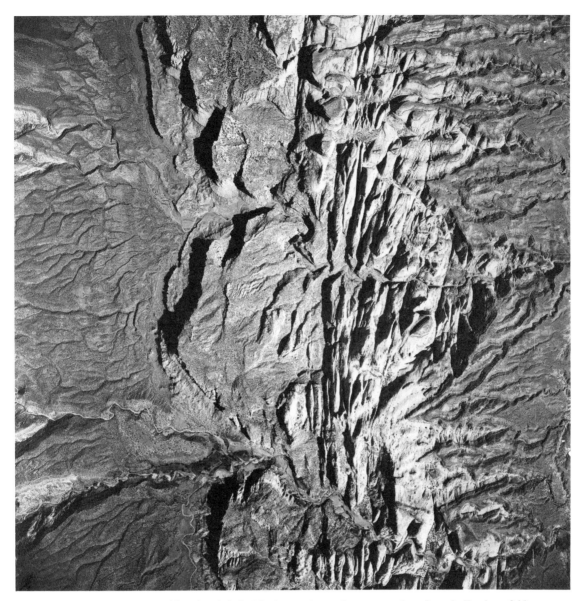

Figure 4.D. Example of fissuring (jointing) in the massive sandstones turned up along the Escalante fold (= Capitol Reef). The massive white sandstone is the Navajo, capped by Carmel Formation forming the flatirons at the right. The big cliff-forming sandstone left of center is the Wingate; the Kayenta Formation forms the ledges between it and the Navajo. Dark beds rising westward at left are Triassic, Gilbert's Shinarump Group. Underlying white sandstone is top of the Permian. Capitol Gorge crosses center of the picture; Pleasant Creek (Gilbert's Tantalus Creek) crosses the bottom. Compare Figure 9.D. (Photograph by Fairchild Aerial Surveys)

Climbing another hill in which I descend through the A6, A8, A9, and 50 ft. of A10 I have an opportunity to measure the balance of A10 (Fig. 4.21).

Wednesday, September 1. Camp 45, on the Dirty Devil River just below the cañon through the Escalante fold. Camp fire is built of wood cut down by beavers—for what purpose? They cannot build dams here.

25.97 = 4630 = Camp 45

We at once enter the cañon cutting through the foss Jura here '   ' at least a heavy sectile bed. The gypsum appears above it in moderate force.

{p. 43}   Cross lamination dips SE. We linger a long time in the White G.M. on account of the duplication of the fold at this point.

The route is impassable at a high stage of the river. There is no well established trail through the cañon as there is through

Temple Creek Cañon. As in that cañon the gravel is chiefly trachyte and the same shows on benches at the sides.

We cross the D.D. river 88 times and dine where it crosses and is interrupted by the variegated marls [*Chinle Formation*].

The dip is about 10° and increases slightly toward the Carb.—which shows in the cañon of the Roundtop Mesa [*Miners Mountain*].

The base of the Vermilion is unusually definite and the shales below are interrupted by but a single sandstone above the variegated.

The top of the Roundtop Mesa is the yellow beds above the cherty.

Dinner 25.78 = 5010

{p. 44}   Marches 7.20–12.20; 2.20–6.20.

After dinner we leave the river and follow a monoclinal valley between Verm. and Aubrey. This valley runs all the way from upper Corrall Creek to No. 119 [*Stack of Wingate Sandstone at SE end of Teasdale Anticline*] curving with an eastward convexity but it is not a line of drainage except for short washes. [*This valley is the one followed by the highway, U-24, west from Fruita, and it was the route of the prehistoric trail.*] Temple Creek and the D.D. river cross it at right angles and the Corrall Creek crosses it twice, reaching the crest of the Roundtop Mesa in this interval [*?*].

The successive arroyos that run parallel to the structure seem to have started each when the base of the shale was along that line and cut too deeply to follow the retiring monoclinal valley. Though if this guess be good I do not see why the side drainage from the cliff has not been equally persistent.

The Shinarump sandstone is wanting at the D.D. river and for 1 mi. S and 2 mi. N It reappears each way in a

{p. 45}   feather edge. At its north recurrence it swells in 1000 ft to 50 ft thickness and at once asserts itself by making a second cliff (which continues only a short distance). In this it is assisted by the var. shales [*Chinle Formation*] which develop a lenticular limestone bed as much as 30 ft thick. The durable *debris* from this holds a butte on the foothills and upon it I am sitting as I write. I can find no trace of fossils.

The sandstones of the lower Chocolate are in this vicinity and from this point westward they make the profiles of the Roundtop Mesa although the cañons cut to the Aubrey. Southward however they are denuded to the very base of the mesa slope.

Toward M.L. (south flank) the dip of the Verm. diminishes very greatly and perhaps it reverses; i.e., the fold runs further N and the cliff stands on its summit.

Going a little farther I find that the prominence of this Shin. spur is aided by a fault—a small fault of only 150 or 200 ft as its throw

{p. 46}   to the N (like the throw of the fold. The trend of the

Figure 4.22. "Section across the fault, p. 46), that accents the ledge of Shinarump. *a* is the Verm. *b* is a thin sandstone above the var. shales [= *Chinle Formation*]. *c* is the limestone in the var. shales. *d* is the Shinarump. The dip of the beds is 5° to the S. The inclination of the fault is 83° and also to the north.

fault is very nearly NW) but it is not straight. My view of it is peculiarly fine (see Fig. 4.22).

Later I find that the fault continues to camp 46, and at one point holds a vein of iron ore. It veins just below a dome of Vermilion near camp 45 and from that I must get its trend for I have made some blunder in my bearings.

Thursday, September 2. Camp 46 = 24.24 = 6150 = 7 am to Supply Camp.

The hill near camp 46 is capped by trachyte debris and along the edge this is carved by drifting sand.

Soon after leaving the fault

{p. 47}   locality last evening I came upon level strata and at this camp they are substantially level. The rise between the Shin. at the N and the D.D. fold at Wheeler's Crag is only a small amount 400–500 ft., say. The change from 10° at yesterday's dinner camp to a merely nominal dip here takes place somewhat abruptly at a point 4 miles E of here. From the point where I first saw the fault to the D.D. fold at the W side of Square Top Butte there runs a monoclinal with westward throw. It changes the N-S section from the full line (Fig. 4.23) (as at Square Top/Roundtop) to the dotted line as at Wheeler's Crag and its max throw is about 750 ft. Its max dip is not great—prob. not 10° and its trend is a little E of N. I do not know whether it extends beyond the limits indicated.

Figure 4.23. Change in section (from full line to dotted line) attributed to monocline (p. 47).

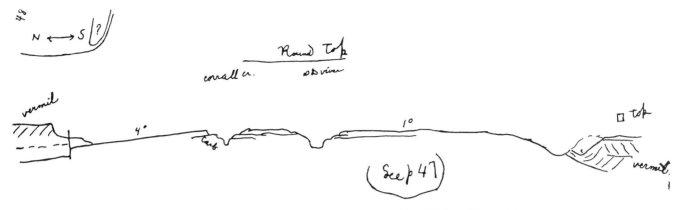

Figure 4.24. Profile and section across Round Top Mesa (= Miners Mountain).

Looking beyond this mono I have in the profile of Round-top Mesa (Fig. 4.24) a true line for a section across the D.D. and Escalante folds, for the whole profile is made of the Chocolate Sandstone. The cañons of both Corral Cr. and the D.D. river must cut well into the Aubrey.

I can see nothing of the limits of the fault. It is a little greater in throw here I think than where

{p. 48}   described yesterday.

I thought I detected in the twilight last night another point of failure of the Shin.

The prevailing wind as indicated by sand etching is from the W.

{p. 49}   **Friday, September 3, 1875.** Supply Camp 8 am = 24.66 = 6230

Marching from 1.20 to 4.30 brought us up to the shoulder of M.L. A little above camp (47) I read 22.20 = 9100.

All the Roundtop Mesa beyond the Dirty Devil and beyond Beaver Dam creek [?] is Carb. on the map. Also a little patch between their junction and the cañon of the D.D. to a point opposite W end of Wheeler's Crag.

From No. 119 to the D.D. cañon the Vermilion holds a straight line of cliffs and for a mile beyond. Then the line turns to the west and it is at the turning point that the fault crosses it in its eastward course. I think the fault turns sharply north and ends before reaching the lower D.D. cañon—the *a* (Fig. 4.25) being turned down like a dog-eared leaf. I am looking WSW at it and this is the view (p. 50) (Fig. 4.26). [*I think Gilbert misrecorded his direction; he seems to have been looking easterly.*]

I have not been able to trace the fault westward as I climbed

{p. 51}   today; it may have run out.

Found a knot in Shinarump wood today.

This point is Graves loc. 29.

Saturday, September 4. Camp 47 = 22.12 = 9200

The same conglomerate limestone that I found before on

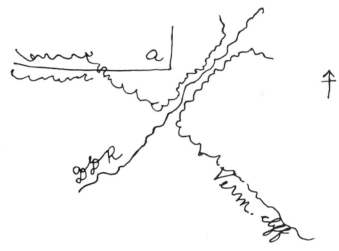

Figure 4.25. Gilbert's sketch map of the bend in the fault NW of the canyon of the Dirty Devil (= Fremont) River.

M.L. I missed this morning—at 21.83 = 9550—a considerable area.

Camp 48 is under the NE corner of M.L. which I climb joining Graves.

Accordant with the nonconformity of Tertiary about M.L. and Aquarius I fancy I can discover the same thing in the Musinia Plateau. Its profile as seen from here is horizontal while the plane of the top of

{p. 52}   No. 4 dips a degree or so to the west. Against this hypothesis is the fact that the surfaces by which I recognize the dip of No. 4 are all nearer than the Ter. and that, in part, the disturbance of Howell's Hole intervenes. In its favor is the fact that the dip of No. 4 corresponds to the general dip of the NW side of the S.R. Swell.

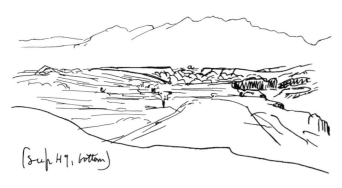

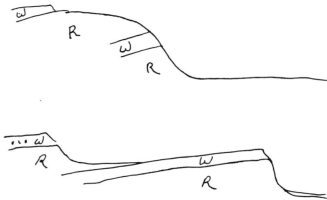

Figure 4.26A. "In this view *a-a'* is the cañon of the river, and the cross section of the fold at *a* seems to be a double mono. like the Kaibab (Fig. 4.26B). The cañon cutting between the branches. At *b* I fancy it thins but it is only a guess (Fig. 4.26C). The throw has all passed to the eastern or northern division except a little that survives in the fault (Fig. 4.26D)."

Figure 4.27. Section of the Belted referred to on p. 52. "In great part the conglomerate is carried back from the crest and in part the two white bands cap separate cliffs.'

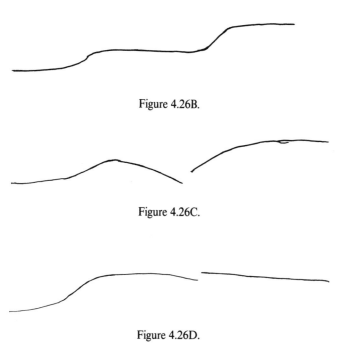

Figure 4.26B.

Figure 4.26C.

Figure 4.26D.

The Belted Cliff here deserves the name, for its middle is belted throughout with white. This white and the red above and below it are Vag. and the white conglomerate on top is not thick (Fig. 4.27).

There is the same confusion about the black top buttes and the minor folds which lie north of here along the base of the trachyte. Must be seen nearer and from

{p. 53}   lower station to be known. The birdseye view is of limited use in the study of structure.

The Belted Cliff is uninterrupted to this base of M.L. The Blue Basin synclinal which it limits maintains its own drainage almost to the trachyte and then is cut across by drainage crossing

in the trachyte a little N of here and leading to the Black Butte anticlinal [*?*] and beyond.

Looking northward toward the Castle Valley region—

The Belted Cliff, after some mangling and perturbation along the lava foot in the foreground, takes up a northerly course. It soon meets with a fold of northerly throw and turns square to the right to follow it for five miles. Then with an acute angle it resumes its northerly course and becomes at once one of the in facing cliffs of the S.R. Swell.

Above the conglomerate of the Belted are white shales at first with just the hint of a hard bed above them.

Then come gray shales, dark and light, capped by heavy, yellow sandstones—heavy but not massive. This combination makes a definite cliff that would correlate with the Kaiparowits if the Henry Mt. region did not throw doubts as to the equivalency of individual sands and shales in the Cret.

The cliff emerges from the trachyte 8 miles N of here— where the spur from Hilgard joins that from M.L. and runs N parallel to the Belted

{p. 54}   except that it is not deflected by the fold. It is plainly crossed by the fold but the increase of erosion to the south that the fold tends to induce is counteracted by the resistance of a trachyte debris capping that extends about to the fold.

The third line of cliffs is a low one—subsidiary to the fourth.

The fourth is made of gray shale covered by twice as much yellow sandstone as pertains to the second. This sandstone is Howell's No. 4. As a cliff it loses characteristic under the trachyte a little this side of windy butte [*?*], but I see no outcrop of what is prob. the same sandstone on the S face of a hill about E from Hilgard.

Sunday, September 5. Camp 48 = 21.61 = 9850

Descending to 22.15 = 9200 I am still above all outcrops of

the fold. The nearest exposure is of Jura gypsum, is 500 ft lower and seems to rise toward me.

Descending to the outcrop I find myself on gypsum at 22.74 (= 8450). If the dip rises toward the M.L. it is slight. The chief and perhaps only dip is to the N in conformity with the fold. A profile near me is as follows (p. 55, Fig. 4.28).

Figure 4.29. Cross section of valley with Belted Cliff (Morrison Formation and gypsum rocks.

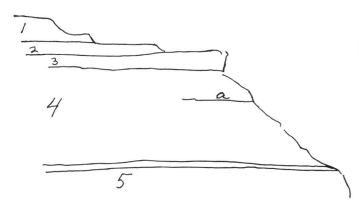

Figure 4.28. "Section and profile of Jura near N. base of M.L. [*Thousand Lake Mountain*].

1 = gypsum and red and white shale
2 = sandstone? (yellow) and yellow shale
3 = yellow sectile
4 = yellow shale interrupted by sandstones and at *a* by a white bed that may be gypsum
5 = G.M. white
No. 1-4 incl. make up about 300 ft."

{p. 55}   Here on the gypsum I find once more the drooping pine and it seems to be strictly confined to the gypsum. Its neighbors are *P. ponderosa* and *A. Douglassii.*

At 23.03 = 8090, while I am below the top of the G.M. on the fold crest, I am still many hundred feet (perhaps 1000) above the top of the Belted Series in the synclinal N of the fold. The max. dip of the fold is here about 10° and is exhibited by the gypsum—not where I have just visited it but at the south margin of the valley faced by the Belted Cliff (Fig. 4.29).

Northward I see in the foreground a syn. running N-S. This is exhibited for a short distance by the Belted rocks but cannot be traced in the sandy flats below. Its W margin makes an anti, with the branch of the Belted which runs northward at the base of the Cret.

What I took yesterday for limestone proves to be merely a saline efflorescence.

{p. 56}   All over the slope this morning I have found the soil to consist largely of a non-volcanic gravel. This must be pre-trachyte and is probably of the same rock series with the

conglomerate-limestone. Since I struck the gypsum this am, I have noted many exposures of white clay and sectile sandstone fragments such as are common in the exposed gypsum beds further south. This would seem to indicate such structures as are exposed above the gypsum and below the Vag. sand in the valley below me are red.

At my lowest point (23.33 = 7740) I still am several hundred feet above the Belted Cliffs on both sides of the synclinal, and command the whole synclinal.

On my way to Dinner Lake, at 22.70 = 8490, I struck conglomerate and sandstone in place. It seemed to have no dip although it would overlie the prolongation of the folded rocks.

Dinner Lake = 22.93 = 8220

We pass other lakes as we travel through the region of slides. The ridge equivalent to M.L. continues to the foot of Hilgard at least.

I have ahead two insulated outcrops that show 'curvalin' [*curvalinearity?*]. The first is red shale with white hard bands, the second yellow sandstone.

{p. 57}   These prove to be—resp—Belted Cliff and Cret. rocks.

Salt Lake near camp 49—23.57 = 7480.

From camp I ride and walk E to the edge of the trachyte (23.46 = 7580) and find myself 150 ft above the lower lead cliff and overlooking a plain of Belted. The section is Figure 4.30.

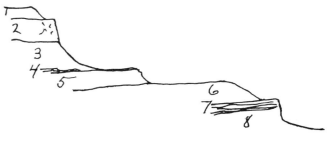

Figure 4.30. Section across *"Gryphaea"* beds near camp 49.

"1 = gray shale protected by trachyte [*debris*]
2 = heavy bedded yellow sand with coal
3 = gray shale
4 = a thin sandstone (place of *"Grayphaea"*) yellow
5 = white or ashen shale
6 = red and purple shale with some white sandstone or conglomerate
7 = a white bed, the hardest in the Belted Series
8 = red belted (total 2000?)"

These are all dipping to the W (as far as the crossfold), the ridge at 6 with an inclination of 10° and that at 2 with 8°(?). Beyond the cross fold they are almost level. I cannot trace the crossfold further than before. The measure of its (N) throw is about beds 4–8 incl.

{p. 58}   There is a snow bank on E side of M.L. During our shower today M.L. was whitened.

See notebook No. 5

Letter list

July	8	no. 25 - home
July	11	no. 26 - home
July	15	no. 27 - Dutton
Aug.	10	no. 28 - home EEH

{p. 61}   [*These pages reversed; recorded from back forward.*]

Altitudes corresponding to bar inches--uncorrected

Inches	Feet	
18	14,000	
		--140 for 1/10 inch
19	12,600	
		--135
20	11,250	
		--130
21	9,950	
		--125
22	8,700	
		--120
23	7,500	
		--115
24	6,350	
		--115
25	5,200	
		--110
26	4,100	

[*The following set of numbers evidently is calibrating three barometers.*]

	444	Dt	3		9	
21 pm	24.307	78	23.990	309		
22 pm	24.299	77	24.050	249	23.610	689
23 am	24.307	71	23.940	367	23.590	817
Ells	(22.330)		22.080	(245)	21.580	(750)
24 am	23.955		25.170	1.215		
24 pm	24.029	74	25.120	1.091	23.310	719

Aneroid 3 changed 1 1/3 inches descending Ellsworth on the 23ᵈ pm.

The distance I need for estimate of the thickness of a bed of Vermilion on the Escalante fold is ⅞ mile.

{p. 60}   *Route Notes*

Left Supply Camp August 12. Stopped for dinner on creek at the SE base of Wheeler's Crag. Then passed to the left of

Square Top butte and continuing in the same direction encamped on a small short creek within 2 miles of Temple Creek.

Aug. 13. Followed Temple Creek through its two cañons.

Aug. 14. From Temple Creek to Pleasant Creek wash. Then westward again to a wash cutting the great fold close to No. 119. Dinner. Then southward across Pleasant Cr. Wash to water-pockets of Thompson's trail at the foot of the fold.

Aug. 15. Camp held. Graves, Adams, and I ascended Tit Mesa.

Aug. 16. Moved to pockets near the head of Waterpocket Cañon. Graves made Sta. 23.

Aug. 17. Marched down Waterpocket Cañon.

Aug. 18. Marched down Waterpocket Cañon and encamped just below its Horseshoe.

Aug. 19. Camp held. Bell, Graves, Sorenson, and I climbed Point XI and were out all night.

Aug. 20. Marched up Waterpocket Cañon to Adams Pocket. Hillers and Jerry detached here.

Aug. 21. From Adams Pocket to Mt. Hillers. Followed a trail under the Middle Leaden Cliff. Lunched on Pine Alcove Cr. Encamped on the S shoulder

{p. 59}   of Mt. Hillers. Left Dinah on Piue Alcove Cr.

Aug. 22. **Sunday.** Camp held.

Aug. 23. Camp moved 3 miles E. Graves, Adams, Sorenson, and I climbed Mt. Ellsworth and made day camp at its base.

Aug. 24. Marched around Mt. Hillers on its east shoulder and encamped on its NE flank.

Aug. 25. Camp moved to NE flank of Mt. Pennell. Graves ascended Pennell, Adams and I made an excursion to E base of Henry I and returning encamped on E base of Pennell on bench of Trachyte Creek.

Aug. 26. Adams and I rejoined the party. Graves, Bell and Sorenson climbed Mt. Pennell. Camp held.

Aug. 27. Bell, Sorenson, and Adams detached this morning to rendezvous at Supply Camp. They held camp and we marched, first to the saddle between Mt. Pennell and Henry I [*Mt. Ellen*] and then via crest of Henry I to a spring on the E side 1 mile S of the conical summit (Sta. 27).

Aug. 28. Graves and I climbed Sta. 27 = the North, conical summit of Henry I.

Aug. 29. Descended Henry I on the W side and followed a wash nearly to the Dirty Devil River below Twin Mesa.

Aug. 30. Crossed the Dirty D. River and encamped between the Twin Mesas. Graves and I climbed the North Twin

{p. 58}   making Sta. 28.

Aug. 31. Crossed the Red Arena and encamped on the D.D. River just below its Trias Cañon.

Sept. 1. Ascended the Trias Cañon of the Dirty D. River, dined at its head and then followed Trail valley to Corrall Creek.

Sept. 2. Corral Creek to Supply Camp.

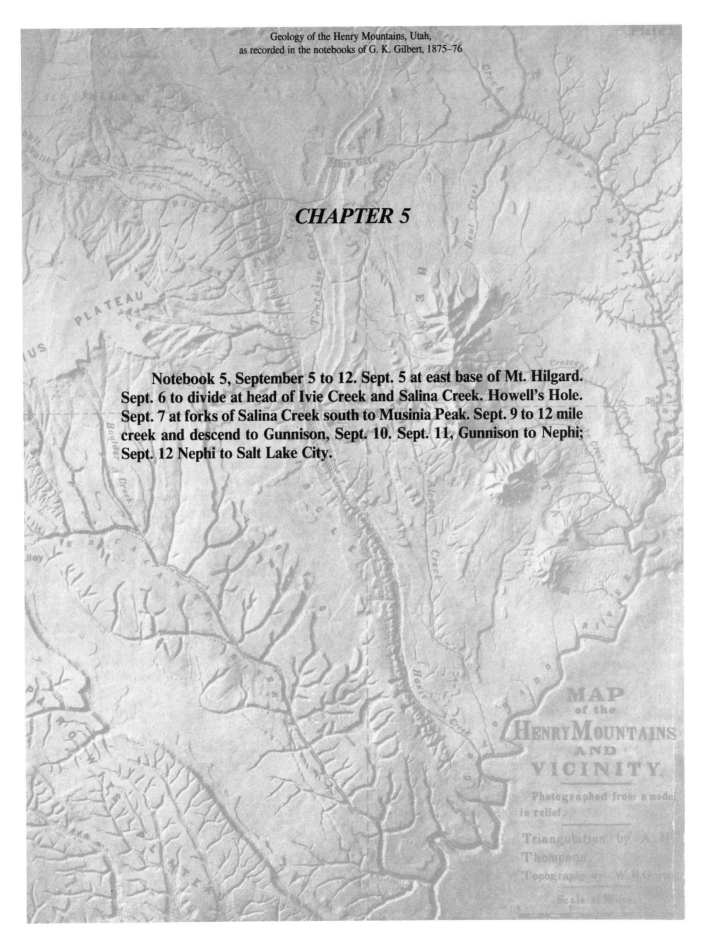

CHAPTER 5

Notebook 5, September 5 to 12. Sept. 5 at east base of Mt. Hilgard. Sept. 6 to divide at head of Ivie Creek and Salina Creek. Howell's Hole. Sept. 7 at forks of Salina Creek south to Musinia Peak. Sept. 9 to 12 mile creek and descend to Gunnison, Sept. 10. Sept. 11, Gunnison to Nephi; Sept. 12 Nephi to Salt Lake City.

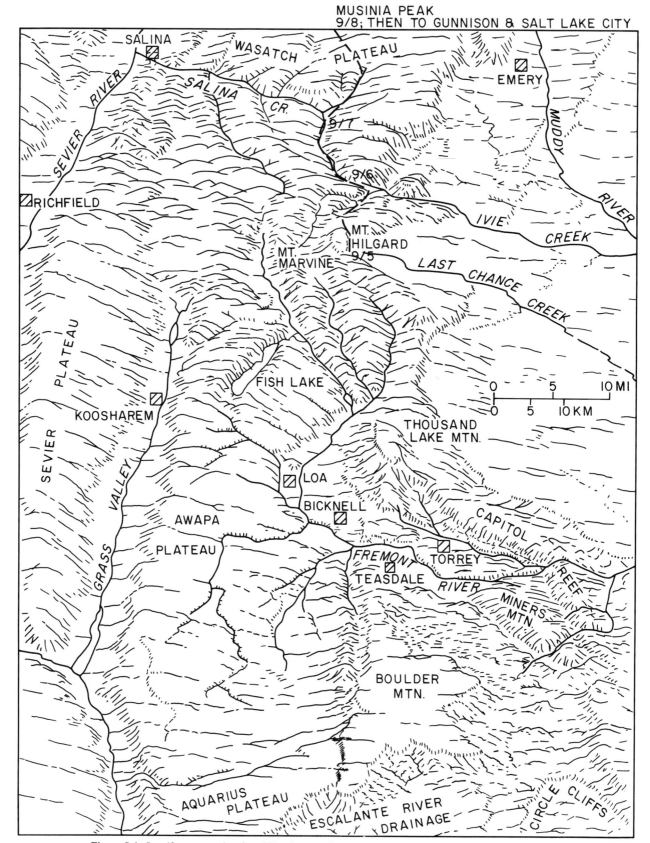

Figure 5.A. Landform map showing Gilbert's route from Mt. Hilgard to Musinia Peak, September 5 to 8, 1875.

Figure 5.B. Air view of Mt. Hilgard. View is NW; the lava cap dips south.

G.K. GILBERT, POWELL SURVEY, BOX 806, WASHINGTON, D.C.
NOTEBOOK NO. 5
OPENED ON THE EAST BASE OF MT. HILGARD, UTAH, SEPTEMBER 5, 1875.

{p. 1}   **Sunday, September 5, 1875.** (continued from notebook 4, p. 58, top)

The rocks which dip west along the cliff front change behind the cliff to an E dip. Northwest, on both sides of the salt lake, the dip is west. And eastward under the cliff the Belted exhibits two changes of dip. So there are two anticlinal changes and two synclinal in a distance of less than 10 miles (Fig. 5.1). The curved strata seen on the march yesterday belong doubtless to the same system. That these displacements belong to the epoch subsequent to the trachyte would seem to be indicated by the presence of trachyte debris beyond (E of) the mono or faulted valley in which the lake lies. In one place the section is thus (Fig. 5.2).

{p. 2}   I read (tomorrow) at the lake 23.75 = 7260; on the divide, 7360.

These displacements and those of Howell's Hole are the most easterly of the Sevier group—a group perpetrated after the trachyte and after the Tertiary and characterized by an abundance of faults. The more ancient system, of which Escalante and San Rafael are examples, shows fewer faults and shows flexure even in the Henry Mts.

By the way, how do the flexures of Ellsworth and Henry V and Navajo consist with the violence belonging to irruption and eruption? Can deeply buried strata be bent without the slowness of action necessary at the surface?

Figure 5.1. The two synclinals and anticlinals east of the cliffs east of Mt. Hilgard referred to by Gilbert on p. 2 of Notebook 5.

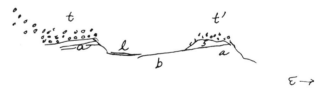

Figure 5.2. Section illustrating gravels isolated by erosion east and west of the lake referred to on p. 2. "l = the lake; t-t′ trachyte debris; s = shale; *a a* = sandstone; *b* = sandstone without trachyte cover except scattered boulders. As things lie now there is lacking a bridge from the transportation of t′ across the lake valley; nor is there a drainage system competent to have moved the sedimentaries between t and t′. The lake has no surface drainage and the lowest divide of its basin is 100 ft above its level."

Monday, September 6. From White Lake to Windy Pt. Ivie Creek (dinner) and Salina Cr. White Lake = 23.75 = 7260. Windy Point = 22.46 = 8780.

From the lake to Windy we have passed many outcrops of sandstone with no great dip. They are obscured by trachyte but not too much to yield their system to thorough study. There is an outcrop on the S side of this {p. 3}   hill, not more than 500 ft. down. I am pretty nearly on line with the fold of northerly throw designated "cross fold" in yesterday's notes (Book No. 4); and I think this upper head cliff (Howell's No. 4) shows in the movement. For on the slope to the N the strata appear to rise toward me and beyond no sandstones seem as high as in this hill.

I think now that the sandstone of the divide we shall cross this pm is a higher bed than the one which rises out of the cliff overlooking Castle Valley. It was the occasion of doubt when I crossed it before.

The red and white exposure nearer and at the base of Hilgard must be Tertiary.

A barometric station on the Musinia Plateau and another in Castle Valley would make a fine couple for an hourly scientific set. At the same time a set of vertical reciprocal angles would test refraction and check the determination of altitudes.

Dinner camp on Ivie Cr—22.53 = 7500 which is Bar. Sta. 2 pm 23.62 = 7400

Forks Ivie Cr. 3 pm, 24.14 = 6820

Basin rim = 23.77 = 7340; camp 50 = 24.62 = 6270

{p. 4}   **Tuesday, September 7, 1875.** Camp 50 is at the forks of the Salina Creek—just above Howell Hole.

From the Divide (Basin Rim) we descended yesterday, first in a cañon entirely in Howell's No. 4; then with the W wall Ter. and the E No. 4; then (after entering the creek valley) with both walls Ter. At camp we are on Ter. barely W of the fault. The line of fault is very plain where we passed it in the creek.

Camp 50 = 24.68 = 6200

The N fork cañon close to camp ascending the S wall of the cañon I read at the top of the first bench 24.44 = 6480; and at the top of the second 24.25 = 6690. These benches are determined by heavy sandstones (yellow). The profile across the cañon is (Fig. 5.3).

The 3d sandstone does not appear in the hill I am on.

Walking 20 rods to the S end of the hill, I find it acute and that its outcrop is due to the tapering of the block which constitutes it. Parallel to the cañon of the N fork and ½ mi. south the block is cut by another cañon due to a drainage from the east. The plan is

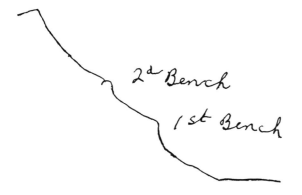

Figure 5.3. Profile of south wall of Salina Canyon near camp 50.

{p. 6}   roughly this (Fig. 5.5).

From *t t* it looks as though this *t t* block dipped to the south a little and from the other end of the hill it seems to dip N.

Between the *t t* and *a a* are hills that I am not yet sure of. S of *B* they are higher than *t t* (a very little) and covered by bushes.

{p. 7}   N of *B* and thence to the N fork valley they are low and seem to hold sandstone. The *B* drainage which cañons in *a a* and *t t* has an open valley in *b b*!

Call *M* the direction of Musinia then the trend of the west fault of *t t* is M 5° W.

M = 335^{aa}; M 5° W = 330^{aa}

a a is (geol) 800 ft higher than *t t.b* is at least 500 ft lower than *t t.*

Returning to the N end of the hill I find the course of the N fork above the cañon to be straight for a mile (or ¾) following the fault between *b b* on the east and *t t* and *d d* on the west. The course of the fault is 10^{aa} = M35°E (see p. 6 and 8).

The sketch on p. 8 (Fig. 5.6) includes blocks of 4 levels—*d d, b b, t t,* and *a a.* If Musinia Plateau conforms to *a a* then the difference bet *a a* and *d d* is something immense if also the rocks of *d d* equal those of Musinia.

In a general way, I suppose the divide block, east of me *consists with* (to coin a phrase) Musinia Plateau; and the Salina Cañon block west of me "consists with" Musinia Peak, and the intervening area of fracture and down throw continues the valley, evidently structural, between the peak and the plateau.

The hill *d d*

{p. 8}   has white at top (dipping toward the Plat.) and is variegated below (Fig. 5.6).

Top of *b b* = 24.23 = 6800 11 am. This is 50 ft lower than the

{p. 9}   rock on top. *b b* is sandstone as shown by small pieces and by a soil of sand.

The fault between *t t* and *d d* passes through x (Fig. 5.6) and is parallel to that on the opposite side of *t t,* i.e., it trends M5°W.

The exposures on *b b* are only along washes. Descending to the creek I find only sandstone and shale, the sandstone pale yellow and very soft, the shale gray. Their alternation is rapid. I take this series to belong to No. 4 somewhere above either of the benches of *t t.*

Creek—24.53 = 6370

Wednesday, September 8. Camp 50 = 24.52 = 6400

The sandstone of the west wall of Howell's Hole is feebly anticlinal at the Salina Cañon—not by a curve but by an angle.

It is now evident that the sandstone of *t t* and that of the Salina Cañon block converge northward but I cannot yet tell how much. The latter's bend is not quite straight but slightly concave eastward. It is about 166^{aa} = 344^{aa} which would give a convergence with *t t* of only 14° + a few degrees given by the curvature.

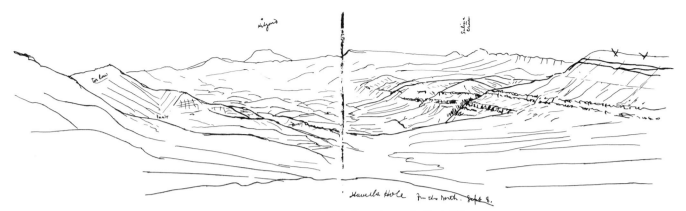

Figure 5.4. "Howell's Hole from the north. Sept. 8."

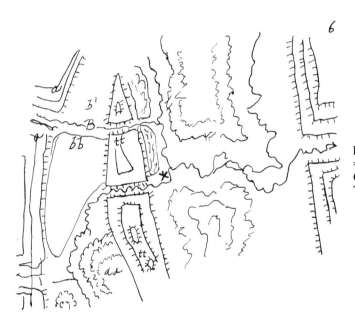

Figure 5.5. Plan of drainage near camp 50. tt = tree top station. The star = camp 50. *a a* = the block which southward makes the divide where Gilbert crossed. The straight lines mark No. 4, and the curved mark soft Tertiary.

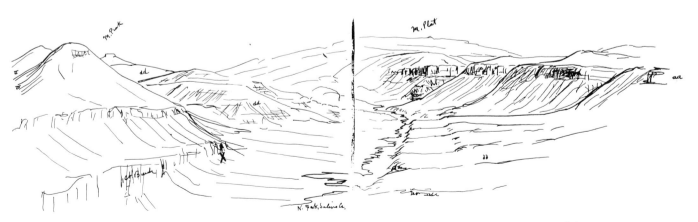

Figure 5.6. This sketch looking north from Salina Canyon to Musinia Peak "includes blocks of 4 levels—*d d, b b, t t,* and *a a.* If Musinia Plateau conforms to *a a* then the difference between *a a* and *d d* is something immense if also the rocks of *d d* equal those of Musinia."

When I reach the Howell's Hole it appears unquestionable that

{p. 11}   No. 4 of *a a* runs consistently under Mus. Plat.; that the fault which limits it runs northward to the W tangent of that plateau and that other faults cross the area obliquely from SE to NW making an angle of 15°–20° with the trends of the great limiting faults. *t t* runs into the Salina Cañon block cutting out the Howell's Hole block completely, and I am not far from the junction. The third bench of sandstone is as high as 1 and 2 combined I think (Fig. 5.7).

Trend of fault *a b* = 146ᵃᵃ at camp 50; fault *h i* = 159ᵃᵃ at Salina Cañon. *u u* has about the same level at junction as *t t* but rises toward *i i. c c* is lower than *a a* and higher than *d d* and dips northward 6°.

{p. 12}   As I ascend, the strata rise with me and I hold about the same horizon to 23.00 = 8120. Then I begin to strike gray calcareous shales and at 22.72 = 8460 I touch a gray limestone (not in place). The beds in neighboring spurs are tinted as are the lower beds of blocks *h h* and *d d.*

Camp 51. South flank of Musinia Peak 22.65 = 8520.

The camp is in a region of slides. The peak (Fig. 5.9) has three terraces. The upper is the peak itself and is due to a limestone bed over shale.

{p. 13}   **Thursday, September 9, 1875.**

Camp 51.   22.65 = 8560. Base of lower limestone 22.20 = 9090. Top of limestone shoulder 21.79 = 9610. Figure 5.10 illustrates the limestone shoulder.

How come the rocks of Howell's Hole have lost this limestone cap? It must never have existed or have been eroded before the faulting. To have been selected for erosion the limestone must have been thin—a hypothesis not improbable considering the variability of the Tertiary section.

The Musinia zone of faults extends through Howell's Hole and the Hilgard region to Rabbit Valley and Aquarius pass, with an eastern branch at White Lake, and everything west of that line is affected in the same way. The prevailing westerly dip that is exhibited bet. Musinia and the Sevier valley is carried southward to this end of Grass Valley including

{p. 14}   Gilson's Crest and Craggy head.

This view throws no new light on the structure of Howell's Hole but everything tends to confirm the views already noted.

Beside the land slip lakes noted yesterday there are two north of M. Peak and one east. I see another N of 12 mi. creek.

The amphitheater of 12 mile creek may be structural but I cannot so make it out. It is so full of slides that it would be hard to study.

The cliff beyond this amphitheater (N) and the continuation of the same around to the S tangent of M. Plat. are as high as this

Figure 5.7. Sketch of the benches of sandstone referred to by Gilbert on p. 11.

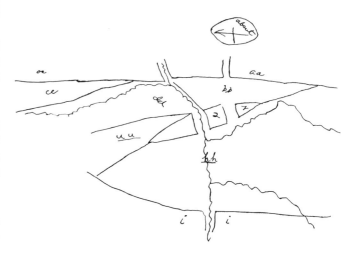

Figure 5.8. Map on back of p. 11. "*h h* dips southward 3°. *d d* dips S. 1° or 2°. I can trace it 6 miles. The star = 23.65 = 7960 = top of 3ᵈ bench of sandstone. This closely resembles the sandstone of *b b*, p. 8, 9."

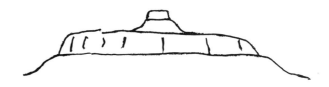

Figure 5.9. Musinia Peak consists of three terraces. "The upper is the peak itself and is due to a limestone bed over shale. The second is equally regular and is due to a heavier limestone bed. The third is uneven and is a shoulder of slides, like the shoulder of Aquarius."

Figure 5.10. "Top of limestone shoulder = 21.79 = 9010. The limestone shoulder has *c* = conglomerate; *u* = uneven; and *l* = limestone, gray and 'freipaund'. Then uneven to foot of climb = 21.55 = 9900. White limestone to summit of *M*. P = 21.36 = 10,150."

Figure 5.11. One of the N.S. displacements with slight throw on Mt. Musinia is shown by the convergence of the top of the upper limestone (*a-b*), which is level, and the top of the lower limestone (*a'-c*) on Fig. 5.11 which is inclined to the north.

Figure 5.12. Arching of the plateau and possible small fault near the great fault of Howell's section. Gilbert Notebook 5, p. 15.

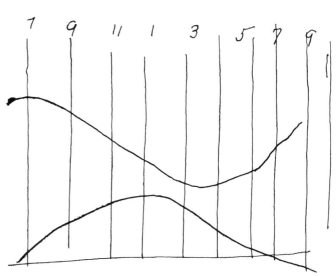

Figure 5.13. Note at Camp 52 on Twelve Mile Creek. "While the temperature curve is concave upward, the horary curve must be descending, while it is concave ascending (?).

"The horary curve is highest when the increase of rate of expansion is most rapid, hence, when the temperature curve is concave upward and has the smallest radius of curvature (?). Which of these two propositions is correct? Is the horary curve a first, or a second differential of the temperature curve?

"If the second proposition is true, then the radii of curvature of the two curves change signs through (∞) at the same time.

"In any case, the horary curve of the barometer is the true datum from which to determine the mean temperature of the air when used in hypsometry."

point and cut off all distant view. There are displacements on M. Plat. of N and S trend and slight throw (Fig. 5.11).

There is no difficulty in ascending the first shoulder of M. Plat. from this side. The upper bench is not so accessible but I think animals can be taken up

{p. 15}   it.

Bet. 12 mile and Willow Creeks there is an E–W displacement but I cannot make out whether a simple fault with N throw or a double fault, the intervening block having dropped. It appears in rocks dipping strongly to the west.

It is possible that some of the surface I ascribe to slide is morainal; but there are surely no glacial phenomena of note.

On the upper shoulder of M. Peak is this 'pine' collected on Oyster Cr.

In the northward view there is a slight arching of the plateau and perhaps a little fault before the great fault of Howell's section, thus (Fig. 5.12).

The surface of the M. Plateau is more broken than I had imagined and may offer a good field for the study of displacements. It needs however to be pretty thoroughly traversed to insure the identification of the fluctuating beds.

Dinner camp 23.05 = 8080. Two more lakes seen.

Camp 52 on 12 mile creek = 25.16 = 5670.

Jerry Picket, Gunnison, Ut.

Jno. Sorenson, Gunnison

Nathan Adams, Kanab

Monday, September 10. Camp 52 = 25.35 = 5470 at 8.45

Gunnison at noon, 25.96 = 48.30

The section along the N side of 12 mile cañon and especially at the N of the road bet. 12 mile settlement and Gunnison is interesting and deserves study as supplemental to Howell's work.

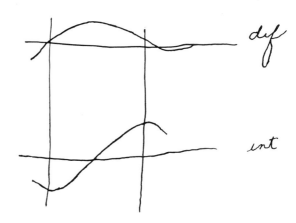

Figure 5.14. "If one curve is a differential of another, then the integral curve will have a max. or min. for every point in which the differential crosses the axis and for no other point."

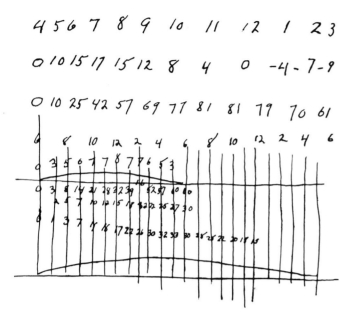

Figure 5.15. "It is probable that the hypsometric (as distinguished from the superficial) maximum of temperature occurs only 2 or 3 hours before sunset—or later than that directly indicated by thermometer."

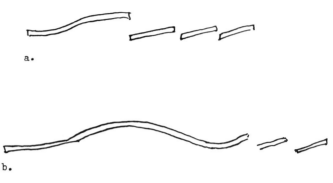

a.

b.

Figure 5.16.a,b Structure section along Twelve Mile Creek. "Instead of *a* as I have drawn it, is *b*."

Almost all the way the Tertiary is to be seen on the eroded edges of the Jura, and the structure section, instead of *a* (Fig. 5.16A) as I have drawn it, is *b* (Fig. 5.16B).

I think the road after leaving San Pete Creek rises through an E–W fault line and that the whole section S of the road bet. the two settlements is higher than north.

Arriving at Gunnison I found all parties ahead of me except Renshaw.

Saturday, September 11. Gunnison to Nephi
Sunday, September 12. Nephi to Salt Lake City

Cash items

Sorenson, Jerry, and eggs	$2.00
Postage in Gunnison	.25
Telegraph in Gunnison	1.00
Sept. 12 Hotel and stage, Salina	5.00
Nephi to Salt Lake, fare and c	5.00
Salary July and Aug.	400.00
Bus in Salt Lake	.50
Hairbrush and comb[1.60] sponge[1.00]	2.60
toothbrush[50] tablets[50]	1.00
hat[3.50] notebook[25]	3.75

{p. 21}   Salt Lake items

clothing	Pop science
hat	telegraph
mail	hair
bath	'Simple Girdle'
toothbrush, sponge comb	
drawers 2	

{p. 23}   **June 27, 1876.** Specific gravities of Utah Rocks
[*These determinations of specific gravities of Utah rocks indicate that Gilbert already was thinking about relative densities as a factor controlling the intrusive mechanism at the Henry Mountains, a matter discussed on pages 75–80 of his monograph.*]
Henry Mt. lava coarse grained. No. 381. string = 130 millegrammes.
Dry 466.650. In H_2O = 288.995.

$$466.650 - 130 = 466.520$$
$$288.995 - 130 = \underline{288.865}$$

$$466.520 \div \qquad 177.695$$
$$= 2.63$$

Trachyte from Salina Creek, No. 351
Dry, 204.600; in H_2O, 125.020

$$204.600 - 130 = \quad 204.470$$
$$125.020 - 130 = \quad \underline{124.890}$$

$$204.470 \div \qquad 79.580$$
$$2.57$$

{p. 24}   Sandstone of "Trias" from Mt. Hillers, 20 ft from dike. No. [*not entered*]
Wet, 310.650; in H_2O = 177.840
Lost a little by disintegration

$$310.650 - 130 = 310.520$$
$$177.840 - 130 = \underline{177.710}$$

$$310.520 \div \qquad 132.810$$
$$= \quad 2.33$$

Slaking shale of Shinarump from M.L.

[*page blank*]

{p. 25} List of the Wheeler notebooks of 1871 and 1873
A. From New York to Silver Cañon, Nev.
B. From Silver Cañon to Desert Wells, Cal.
C. Desert Wells to Camp Mojave, Ariz.
D. Camp Mojave to Catastrophy Rapids in the Grand Cañon.
E. Grand Cañon to Prescott, Ariz.
F. Black Hills, Ariz. to Sacaton on the Gila
1873
L. (12) Washington to Ojo de Benado
M. Colorado Ridge to Camp Apache (last time)
N. Camp Apache to Ralston.
O. Ralston to Sierra Datil.
P. S. Datil to Washington

{p. 26}   Wheeler notebooks of 1871
Letter list
July 8 - No. 25 Home
July 11 - No. 26 Home
July 15 - No. 27 Dutton
Aug. 10 - No. 28 Home - EEH
Sept. 3 - Postals F.P.G. and G.S.G.

{p. 60}

Inches	Feet	
18	14000	
		140
19	12600	
		135
20	11250	
		130
21	9950	
		125
22	8700	
		120
23	7500	
		115
24	6350	
		115
25	5200	
		110
26	4100	

Hillers marches
Aug. 22. Remained at Adams Pocket
Aug. 23. Adams Pocket to mouth of Pleasant Cr. Cañon.
Aug. 24–25. Held camp.
Aug. 26. Pleasant Cr. Cañon, via old Dirty Devil Trail to camp near Mosquito Point.
Aug. 27. Marched to Bowlder Cr. camping in sandstone cañon.
Aug. 28. Moved up to mouth of creek (on which we camped).
Aug. 29. Moved up to junction of Bowlder and Cataract Crs.

{p. 59}
Aug. 30. Held camp.
Aug. 31. Bowlder Cr. to Aspen Lake.
Sept. 1. Aspen Lake to Supply Camp.
     Water supply routes, names, timber, Shinomo traces, and other Indian tracks.
Marches, etc., from Supply Camp to Gunnison
     Sept. 3. Left Supply Camp at 1.10 pm and marched by short cut to the shoulder of 1000 Lake Plateau.
     Sept. 4. Bell, Graves, and Sorenson climbed 1000 Lake at the old trail and descended from the north end. Moved camp to a spring 500 ft below the NE salient.
     Sept. 5. Marched northward along the base of the slope of trachyte debris, encamping in a close valley at the eastern base of Mt. Hilgard.
     Sept. 6. Dined on the S fork of Ivie Creek and encamped at the main forks of Salina Creek. Graves, Jerry, and I stopped a few hours on Windy Peak in the am.
     Sept. 7. Bell, Jerry, and Sorenson marched to the base of Musinia. The remainder of the party remained in camp and fished.
     Sept. 8. [*blank*]
Routes of travel
     In going from Supply Camp or Aspen Lake to the Henry Mts. there are two chief difficulties to be overcome—first, the line of sandstone highlands that Hoxie called "Imprenatable Ridges"; second, the system of cliffs and mesas that lie at the western base of the Henries.

Top upper lead cliff	+3200
Top middle lead cliff	+2400
Top lower lead cliff	+ 600
"Gryphaea" bed	0
Top Jura gypsum	–1200
Top fossiliferous Jura	–1400
Top Vermilion Cliff	–2400
Base Vermilion Cliff	–3400
Shinarump conglomerate	–3800
Top Aubrey	<u>–4300</u>
Top Kaiparowits sandstone	+2100

     Rept upon Geog. and Geol. Espl. and Surveys W of the 100th Mer. in charge of 1st Lt. Geo. M. W. etc etc Vol. III Geology

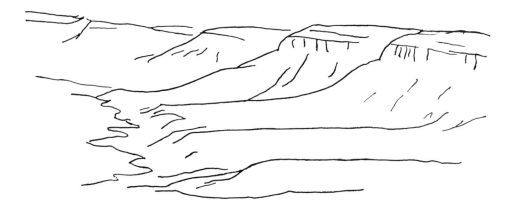

Figure 5.17. Drawing with no caption and without reference in the text in back part of Notebook 5.

CHAPTER 6*

Notebook 6. The first part of this notebook, covering July 17 through August 20, 1876, concerns Gilbert's travel west from Washington, D.C., and a trip to the northern part of the Lake Bonneville Basin (see Hunt, C. B., ed., Pleistocene Lake Bonneville, ancestral Great Salt Lake, as described in the notebooks of G. K. Gilbert, 1875–1880: Brigham Young University Geology Studies, vol. 29, pt. 1, Dec. 1982, 225 p. Pages 23–27 repeated here). Aug. 20–22, York to Nephi and Gunnison. Aug. 23–29 at Gunnison assembling party. Aug. 30 to Salina Bridge. Aug. 31 to Sept. 3., Salina to Rabbit Valley. Sept. 4–6 to Capitol Gorge, route of the prehistoric Indian trail through the Capitol Reef. Sept. 7 to Lower Temple Creek (= Pleasant Creek). Sketch of the Pleiades. Sept. 8 to Tantalus Creek (= Sand Creek) pass, rounding the north end of Stevens Mesa. Sept. 9 to dry camp in badlands northeast of Stevens Mesa. Sept. 10 to Cache Camp on Dugout Creek where it issues from Mt. Ellen. Sept. 11, 12, ascending Mt. Ellen to the peak. Sept. 12, 13 at camp near the peak, making sketches in all directions from there. Sept. 14 to camp at foot of Sawmill Basin, Gilbert's "Bowl." Sept. 15, 16, top of Bull Mountain (Gilbert's Sta. 2). Sept. 17 to Wickiup Ridge and camp at spring in Crescent Creek. At back of this notebook is Gilbert's list of supplies.

---

*Refer to Table 0.1 or appendix for abbreviations, names, etc.

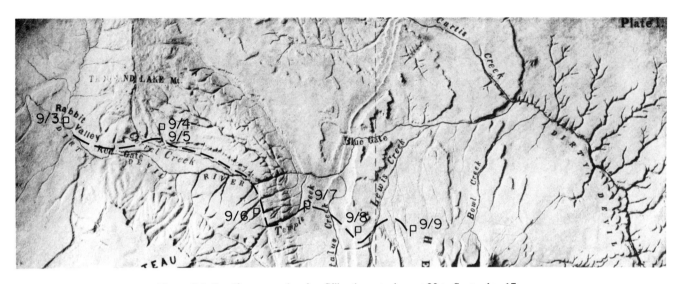

Figure 6.A. Landform map showing Gilbert's route August 20 to September 17.

Figure 6.B. View of the face of the escarpment at Capitol Reef showing the formations described by Gilbert in Figure 6.4 and the distribution of debris on the different slopes he described, near the location of Figure 6.5.

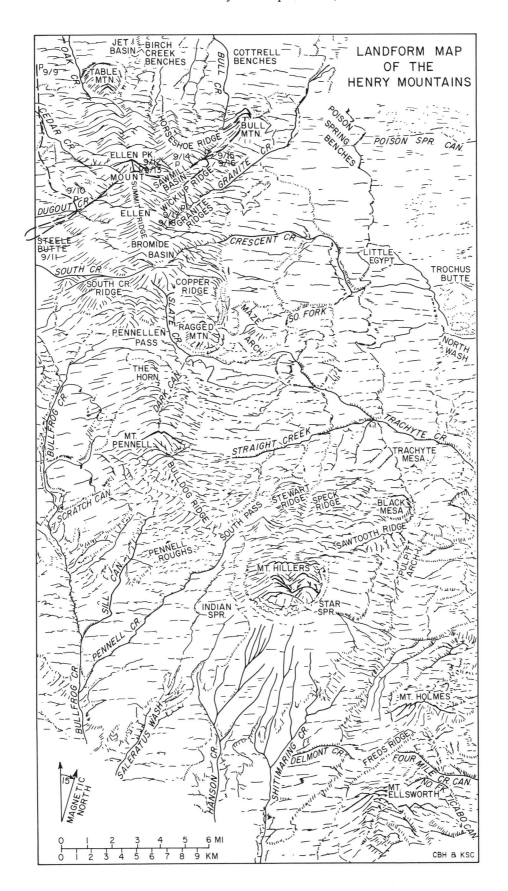

LANDFORM MAP
OF THE
HENRY MOUNTAINS

G.K. GILBERT, POWELL SURVEY, BOX 806,
WASHINGTON, D.C.
**NOTEBOOK NO. 6 [*P. 1–22 CONCERN LAKE*
BONNEVILLE AND ARE OMITTED HERE.]**

{p. 23}   **Sunday, August 20.**

From Salt Lake City to York and then in saddle to Nephi. Nearly all the way from York to Nephi the foot of the mountain shows an appearance that I take to indicate a very recent movement along the great fault that separates the Wasatch (Mt. Nebo) from Juab Valley. Along the limit of the rock at the foot of the Mt. there is a drop valleyward of 10 to 20 or 30 ft that has taken place so recently that the detritable cones are transected and have not obliterated the bench, i.e., there is a fault so recent as to be still marked by a

{p. 24}   topographic bench in soft material. There is no opportunity to compare its antiquity with the Bonneville beach. It seems as well preserved. Surely Mt. Nebo seems to be growing. It is noteworthy that the displacement is uniform or nearly so for ten miles. The masses are moving as one.

Monday, Aug. 21. Nephi to Chicken Creek. On the divide between Nephi and Chicken Creek—10 am = 24.55 = 6360.

20 ft above Chicken Creek pond at 12.30 = 24.63 = 6260.

Plainly the Bonneville flood did not fill Juab Valley so as to connect Sevier drainage with the Jordan and probably it did not even rise above the pond. The Chicken Creek drainage could not have been complicated by the Bonneville flood and one of the problems I came to solve has no basis for enquiry. At 2.15 and 10 ft below the pond 24.78 = 6080. (The discrepancy between this reading and the last is to be explained by the fact that the weather is stormy.)

The pond is artificial, being made for the purpose of controlling the

{p. 26}   water for irrigation. The rock against which the dam rests is a volc. tuff that probably conforms to the strata beyond (west of) it. The rocks all along the west side of Juab valley seem to dip to the east and the tip of the ridge is peculiarly level, N and S.

Along the east side of the valley there is probably a recent fault just as along the base of Nebo. From the fact that there are some points at which the bottom of the valley is 'carried' nearly to the east margin I surmise that the block to the west of the valley extends under it and is depressed along the faultline (Fig. 6.1A).

At 5.15 pm I read 24.40 = 6520 on the hill a. The rock looks like Ter. limestone. I command the Sevier Basin above cañon and see no beaches.

{p. 28}   **Tuesday, August 22.**

Chicken Creek to Gunnison. Pass between Juab and Sevier near Taylor's Ranch at 11 = 24.68 = 6200. River floodplain at 1.15 = 24.79 = 6080.

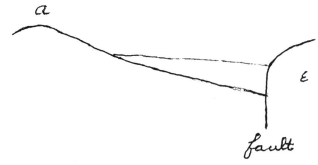

Figure 6.1A. Gilbert's sketch of the inferred relation of Juab Valley to the fault blocks along the east side of the valley.

Figure 6.1B. Cross section, location probably 1000 Lake Fault at Red Gate, near present day Lyman.

"1 = Jura. 'Dippend' S.
2 = White bedded tuff (Ter.?).
3 = green and yellow massive rock, finely fragmental.
4 = porous trachyte? light brown.
5 = basalt breccia cemented by calcite.
6 = dark gray dolomite.
7 = talus cone."

[*Some unexplained compass bearings and some unexplained calculations on this page and the next are omitted here. Pages 29 to 31 are missing.*]

{p. 32}   **Wed. Aug. 23**
               Thurs. Aug. 24
               Fri. Aug. 25
               Sat. Aug. 26   Gunnison
               Sun. Aug. 27
               Mon. Aug. 28
               Tues. Aug. 29
Wednesday, August 30.

Dutton, MacCurdy, and Lewis—Gilbert, Tully, Averitt, and Farnsworth make up the party from Gunnison to Salina Bridge.

Thursday, August 31.

From Salina Bridge to head of King's Meadows. The first is all in Jura-Trias and trachyte debris. The rocks are chiefly green gray shale patched with red and interleaved with gypsum and sandstone. There is no rock resembling the J-T of the Red Desert unless it be that portion above the foss. Jura. Near Sevier Valley is a lime bed dipping west which may be Tertiary.

18.10 at camp on head of King's Meadows = 5.15 pm = 24.317 [77°].

No. 3 = 23.91 = 7080
No. 8 = 24.60
At 5.25 (75°) 18.10 = 24.314
(15.446)

8.868

{p. 34}   **Friday, September 1.** King's Meadows to the Widow's.

Saturday, September 2.
Widow's = 23.08. Summit = 21.50. Fish Lake 21.65
69°   21.986   3 pm.   10 ft above lake.
Sunday, September 3, 1876. Fish Lake to Rabbit Valley. Camping on Nephi Creek.
Monday, September 4, 1876. Nephi Creek to Upper Corral Cr.

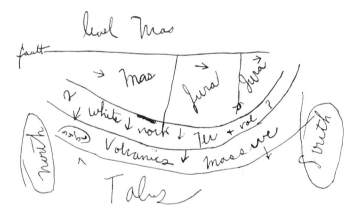

Figure 6.2. Map on page 35. Probably north side of Red Gate. Evidently refers to the cross section in Figure 6.1B.

{p. 35}   The beds in which I found fresh water shells last summer

{p. 36}   seem now to merge in every way with the tuff (2 p. 18) [*Gilbert must be referring to Figs. 6.1B and 6.2; p. 18 concerns his trip in northern Utah.*] and they are wrapped around the blocks of Jura and Trias in a way that the word unconformity does not describe.

On the march the gray mule Louisa rolls down hill with her pack a distance of 50 or 75 ft. The chief damage seems to be a cut and bruise on the thigh and another back of the ear. [*A half century later, also working by pack train in that country, our pack animals had similar falls, and it always seemed to be the one with the crates of eggs or jugs of wine.*]

The Corral Creek Cañon illustrates Powell's explanation

{p. 37}   of the prismatic change of cliffs (Fig. 6.3A).

Just east of camp and on the Vermilion is an old erosion plain veneered with trachyte boulders. It is not less than 1,000 ft above the present base of erosion.

Tuesday, September 5.
Camp held at upper Corral Creek at the Corral. The corral is 220 yards long from *a* to *b* (Fig. 6.3B) and there was a fence across the cañon at *a*—all very rudely

Figure 6.3A. Powell's explanation of prismatic change along a cliff: "So far as the base of the hard Vermilion Sandstone is above the base level of erosion and a little farther (say to *b* in the figure) its disintegration is by sapping and its face is a precipice. But beyond *b* it weathers in grotesque forms with sinuses along planes of bedding and false bedding and joints."

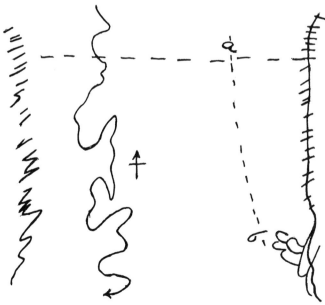

Figure 6.3B. Corral Creek at the corral.

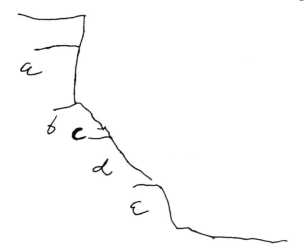

Figure 6.4. "Leaving the D[irty] D[evil] R[iver] the Shin[arump] is absent and the cliff consists of Verm[ilion Sandstone] (a), shale (b), sand (c), var[iegated] shale (d), and choc[olate] shale (e). The debris from (a) is ¾ of it on (b) and (c), ¼ on (d)—almost none reaches the bottom of (e). The sketch slightly exaggerates the slope but it is barely climbable and often beyond the stable angle in (e). Certainly the cap cannot waste more rapidly than it is sapped, but where the cap is trachyte its debris covers the whole slope and endures carriage to great distances." Most of these relationships can be seen in Fig. 6.B.

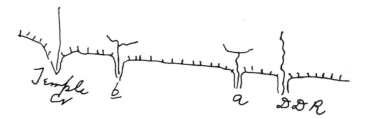

Figure 6.5. Gilbert's sketch map (p. 41) of the position of the two canyons cutting through Capitol Reef between the Dirty Devil River and Temple (= Pleasant) Creek. *b* is Capitol Gorge, route of the prehistoric trail used by Indians to cross the reef and route of the only road connecting Rabbit Valley with Hanksville and other settlements to the east from time of the first vehicles passing there about 1882 until the modern highway was constructed along the Fremont (= Dirty Devil) River (ca. 1960). The National Park Service should have left the old road open as a scenic loop drive for visitors.

a is Grand Wash, tributary to the D.D.R. North is to the right.

Figure 6.6. Rubbed depressions in fallen block that appeared to Gilbert to have been ground out sharpening axes and perhaps chisels. "The former is about 4 inches broad." (Not clear to what Gilbert refers.)

{p. 39}   made and now burned down.

Dutton and I climbed to the shoulder of M.L. and took in the eastward view. Saw elk tracks!

Wednesday, September 6.

From Corral Creek to Dry Camp near Temple Creek, along the monocline. Collected ripple marks with rows of nodes.

The fault noted last year is traceable northward to Corral Cr. The one vein on its plane is of limonite.

At the crossing of the D.D. there are "veneered" plains about 500 feet above present drainage system. They extend so broadly into the mono valley that it seems probable that they represent an epoch when erosion was checked by the lifting of a remote base level or obstruction.

Leaving D.D.R. (Fig. 6.4).

A dry camp 1 mi. from Temple Cr. [*now known as Pleasant Creek*].

{p. 41}   There are two cañons heading in the Carb. and crossing the Verm. between D.D.R. and Temple Creek (Fig. 6.5). The trail turns down *b* (Fig. 6.5) and that is probably the one Hoxie traversed. [*This is Capitol Gorge. It was the natural route through the Capitol Reef in prehistoric times and was the route for the road between Hanksville and Rabbit Valley for 75 years after Gilbert used it. When the highway was built down the canyon of the Fremont River (Gilbert's D.D.R.), the National Park Service closed Capitol Gorge to automobile traffic despite its spectacular scenery and history as a throughway. Visitors have been deprived of an unusually spectacular and historic drive.*]

Thursday, September 7, 1876. From Dry Camp to Lower Temple Creek.

Spent the day sketching etchings and *c* in Temple Cr Cañon. They are all on the NE side and on rock of the Verm. Cl. The chief parts are etched by pounding with a sharp point. In some cases the outline is more deeply cut than the body as though marked out first.

Other marks are produced by rubbing or scraping and still others with colors. The colors are much faded.

I can find no tools, but fragments of pottery and flints and a metate. Several fallen blocks and specially one show rubbed depressions that appear to have been ground out in the sharpening of axes and perhaps chisels. They are of two kinds with profiles like Figure 6.6.

There have been many dates of inscription and each new generation has more or less defaced the work of the previous. So too upon the best protected surfaces as well as on the most exposed there are distinct pictures and others beyond restoration. The period during which the work accumulated was longer by far than the time which has passed since the last. Some fallen blocks cover some etchings and are in turn themselves etched.

{p. 44}   Colors are preserved only where there is almost complete shelter from rain. In two places the holes worn in the rock by tree branches impinge on etchings but the trees themselves have disappeared. Some etchings are left high and dry by a diminishing talus (15–20 ft), but I saw none buried by an increasing talus.

The painted circles are exceedingly accurate and seem to imply a knowledge of the radius.

The Shinarump is not seen from D.D.R. (Fig. 6.4) to *a* (p. 22) [=*41*]. Then it appears intermittently with a wedging to the NE.

The etchings are sketched in the order of their formation from west to east. The numbering merely indicates the grouping given them by interruptions in the available surface. Many were not attempted because too indistinct, others because they were mere repetitions of sheep (mountain?). Some long trails were omitted.

Lower Temple Cr. Camp at 5.30 pm, 24.584 (71°).

{p. 46}   **Friday, September 8.**
From Lower Temple Creek to Tantalus Creek Pass.

This morning Capt. Dutton with Lewis and McCurdy turned back. The Captain had two aqua chills and thought best to return to Rabbit Valley.

The benches at the gap bear trachyte from Aquarius Plat. I count four of which the highest is about 100 ft above water. The stream bears trachyte only in its bed. The flood plain is built entirely of sand.

The Belted [*Morrison and underlying Summerville Formation*] at this place consists of purple and olive gray badland with very little sandstone and conglomerate. The dip is about 10°.

P.M. climbed the lower Blue Cliff north of the gap.

The badlands near the junc. of D.D. and Temple Creek deserve sketching. I do not understand the relation of Temple Creek drainage to Laut Cr. [?] drainage. At one point they both run northward and semi-parallel. Yet the old veneering of Temple Creek seems to extend completely over the Belted Cliff and into the Tantalus drainage. I can only account for this 'seeming on the bypo' that the drainage lines are moving east with the cliffs. Indeed without some

{p. 48}   such yielding the details of drainage would not become (as they do) monoclinal.

The syn. on the Waterpocket swell under M.L. Mt. seems to trend toward this point. This sandstone stands on about the line of max. dip.
Dip =   [*left blank*]
Trend = 330° (2 m) and 160° (2 m).

The Blue Gate Cliff [*Emery Sandstone*] is much higher than the Lower Blue [*Ferron Sandstone*] and its shales are somewhat thicker but its sandstone is little thicker and northward seems even thinner. In general aspects the sandstones are identical and the shales also.

Collected fossils from the *"Gryphaea"* bed. *Loc. a.* is the hardest sandstone (soft enough?) and includes loose specimens of unknown zone. *Loc. B.* is 15 ft above the hardest sand. I fancy the *"Gryphaea"* is a little different from that below.

The shale is arenaceous and yellow green.

The Blue Cliffs are blue when they are

{p. 50}   weathering rapidly, yellow elsewhere, e.g., where there is no cap of sand the color is blue. Where there is notable talus of sandstone fragments the color is yellow. There is more talus on the L.B. [*Ferron Sandstone*] than on the B.G. [*Emery*]. Why?
7 pm. 15 ft above water 69°.   25.150 (16.287) (dif. 8.863).
Saturday, September 9. 15 ft above Tant. Creek at the pass 41°. 25.184 (16.319) (dif. 8.865).
Camp at 8 a.m. = 24.67 (6210).
On the pass through the L.G. Cliff 9 a.m. [*Ferron Sandstone*] 24.34 = 6590

I have seen no trachyte this side of the present Aquarius drainage. The boulders on the Belted badlands are corroded.

Noon camp 12.30, 24.74 = 6150. We meet Henry Mt. trap where we find water on Bell Creek [*probably Sweetwater Creek*].

{p. 52}   The rocks descend almost as rapidly as the arroyo. We follow this creek completely through the L.G. Plateau Wash (which we followed up) below its cañon. 24.63 = 2.45 p.m.

From camp which is in a dry gulch NW of Ellen the strata are substantially level. The North Battery [*Table Mountain?*] shows gray (shale?) and still looks like an 'uppush' like Hillers.

A chapter of accidents. Frank kicked by Little Nephi in the shin. Lightfoot about played out and down twice. My pack bucked off and 3 alfogas [*large, heavy bag for packsaddle; may refer to the straps on the bag*] torn. Evening spent in repairs. Water in pockets bad.
Sunday, September 10. From Dry Cañon NW of Ellen and in the badlands to Cache Camp [*on Dugout Creek*].
Morning camp 6:30 a.m.   24.46 = 6450.

{p. 54}   We find that we followed up the wash next east of the one Graves and I followed down last year. The former carries trap; this latter only sandstone (or almost no trap). The rocks (sandstone, shale) rise toward Ellen (10°) at the heads of the washes.

We have to leave first the horse Lightfoot behind and then the Bald face mule. The latter is brought in in the p.m. The horse is to be sought in the morning.
Monday, September 11. Camp Cache (No. 1) 7 a.m. 23.02 = 8100
Foot of (Sta 28) Butte [*An 1875 station number; Steele Butte?*]. 9 a.m. 23.56 = 7480
Shale to 23.34 = 7730. Then ten feet of sandstone and more shale to 23.40 = 7780 = 9.15 a.m.

Pale buff or yellow green sandstone, friable. Heavy bedded to massive with many grains of a soft white material. 23.07 = 8050 = 9.35 a.m.

10 ft shale
30 ft sandstone
15 ft shale
 5 ft sandstone
Top = 22.98 = 8160 = 9.45 a.m.

{p. 56}   This is the summit of the upper Lead Cliff [*Mesa-verde Formation*] and it is evident that here in the close neighborhood of Mt. Ellen the upper sand is greatly thicker than the middle [*Emery*] or lower [*Ferron*] and can thereby be distinguished. Moreover the middle sandstone is unusually thin here and the lower is thinner than at the pass of the Tantalus. The upper sand is 350 ft and the others 50 ft more or less.

This butte is preserved in the quasi synclinal between the gentle rise toward Ellen and the abrupt rise, thus (Fig. 6.8).

On the south end of the north 'mass' of the butte is a patch of gravel consisting of trap and altered shale. Another conceals the Ellen-Pennell divide and possibly forms the divide bet. South Cache Cr. and Pine Alcove.

There are two and perhaps three of these buttes in the same synclinal case with this. The trend of the synclinal is 140° (4 m) and 340° (4 m). The dip of the B.G. [*Emery Sandstone*] (3 m. N) is 15°.

On next page is the local topography of the streams (Fig. 6.9). The fork of Cache Creek is 2–3 miles above the point where the

{p. 58}   trail leaves.
Descending
Summit = 11 a.m. = 23.020 = 8110
Base main sandstone 11.10, 23.18 = 7800
Horse, 11.20   23.57 = 7450
P.M. marched up Cache Creek Cañon to an opening above the narrows.
      Camp Gomas (No. 2).
Tuesday, September 12. From Camp Gomas (2) over the divide of Ellen to Summit Camp (3).

The valley at Camp two [*probably Ellen Springs*] is occupied by coarse drift that *seems* morainal (Fig. 6.10). [*Not sure what Gilbert saw here; the deposits, a hillside of boulders, are not morainal.*] Climbed Ellen Peak = Sta. 1 = pt x of Graves. Erected flag at 2.40 p.m.

{p. 64}   The entire mass of the main spur (2) (Fig. 6.12) seems to be trap but it lies in talus at the stable slope and may cover something else. In the saddle there may be slate and in the north spur there is certainly slate.

The Hilloid Butte [*Table Mountain*] has much trap and much slate and will bear study. It looks dikelike.

Spur 5 (like 2) shows chiefly trap but it is flecked with the gray color of slate, especially at its western base and the saddle is of a distinct material. The summit of Ellen is all trap. It seems impossible to analyze 2 and 5 on account of the equality of

Figure 6.7. "The Pleides as seen on the morning of September 7 and sketched by moonlight."

Figure 6.8. This butte, sketched by Gilbert on p. 56, but not labeled, probably is Steele Butte.

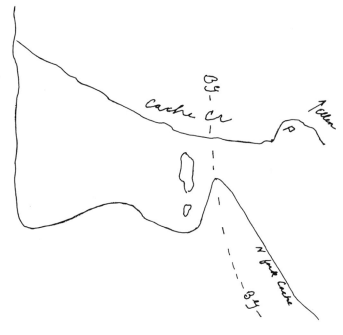

Figure 6.9. Cache Creek = modern Dugout Creek; N (should be S) Fork Cache = modern South Creek; it is labeled S Fork in later drawings. Both are tributary to Sweetwater Creek, which drains north at the left.

Figure 6.10. "Coarse drift" that Gilbert thought might be morainal (p. 58).

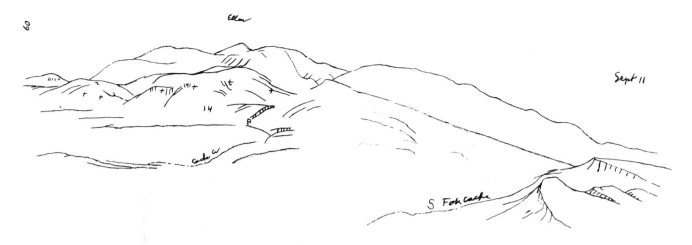

Figure 6.11. Sketch September 11, west side and summit of Mt. Ellen. Cache Creek = Dugout Creek; S Fork Cache = South Creek.

Figure 6.12. September 12. View NW from Sta. 1 = Ellen Peak. 2a, b, c, and d are on the Northwest Spur of Mt. Ellen; 7 is Table Mountain; 6a and 6b are along Oak Creek (designated Cottonwood Creek on the U.S.G.S. Mt. Ellen Quadrangle); 5 is the North Spur of Mt. Ellen (Dry Lakes Peak on the quadrangle map) on Fig. 6.13.

hardness of the components and the masking of limbs and especially talus.

Wednesday, September 13.

Summit camp held. Camp at 7 am, 20.47 = 11,310. Strangers Monument 8.15.   19.54 = 12,570. Saddle under Ellen 8.45, 19.78 = 12,250.

In the saddle are slate and sandstone.

Mt. Ellen summit (aner. 3).   9   a.m. = 19.51 = 12,610

11.30 a.m. = 19.49 = 12,620

2.15 p.m. = 19.48

3.10 p.m. = 19.51

{p. 66}   We can see from here today not merely the La Sal and Abajo groups but two others between them and more distant—also two distinct ranges to the right of Abajo. The Book Cliffs and Navajo Mt. and the Pink Cliffs are comparatively near. We see also Kaibab, Belknap, Monroe Mt., Glenwood Mt., Musinia, Hilgard, Marvine, and finally, a table beyond Book Cliffs [Roan Cliffs]

Taking the middle of the group and not the highest points, some of the bearings are: La Sal—78°, Distant Mts.—140°, Very distant Mts.—101°, More distant mts.—146°, Abajo—109°

[Bearings to La Sal and Abajo Mountains are 5 to 10 degrees too great, depending upon the point sighted. Present declination is about 16.5 degrees; what was Gilbert using? See the discussion on surveying in introductory chapter.]

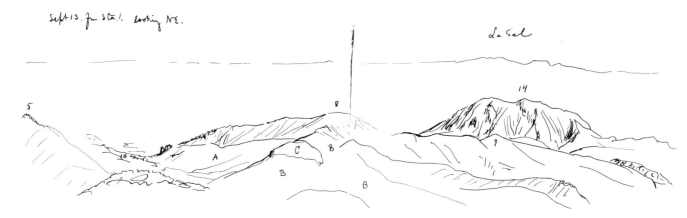

Sept. 13. fm Sta. 1. looking NE.

La Sal

Figure 6.13. "September 13 from Sta. 1 looking NE." 5, see Figure 6.12; 10 is head of Nazer Creek of U.S.G.S. Prof. Paper 228, shown as McClellan Wash on the quadrangle map; 8 is Horseshoe Ridge; 9 is east peak of that ridge; 14 is Bull Mountain, which is Northeast Butte in Gilbert's notebooks and Jukes Butte in his published report. The sketch also shows the La Sal Mountains on the skyline beyond Bull Mountain.

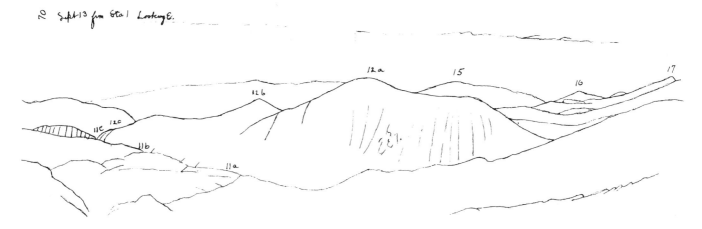

Sept 13 fm Sta 1 Looking E.

Figure 6.14. "September 13 from Sta. 1 looking E." 12a, 12b, 12c = Wickiup Ridge. 15, 16, and 17 are along the Granite Ridge. 11a, 11b, and 11c are along Bull Creek in Sawmill Basin.

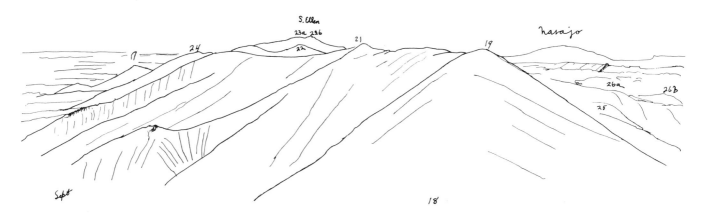

S. Ellen

navajo

Sept

Figure 6.15. September 13. View south from Ellen Peak. 22, 23, 24 are west and north rim of the Bromide Basin and are formed of the shatter zone surrounding the stock there. 17 may be Barton Peak (cf. Fig. 6.14). Navajo Mountain is about 75 miles away. To the right and below Navajo Mountain is the Waterpocket Fold. 19 and 21 are along the North Summit Ridge of Mt. Ellen. 25 is Corral Point. 26a is Durfey Butte. 26b uncertain. 18 is saddle between Ellen Peak and the North Summit Ridge.

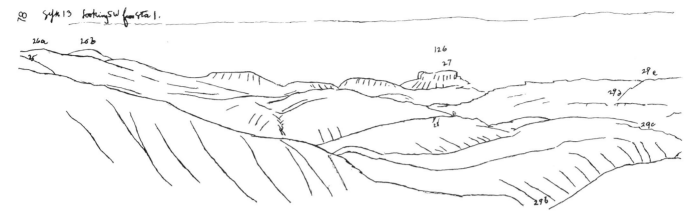

Figure 6.16. "September 13. Looking to W from Sta. 1." 25, Corral Point; 26a, Durfey Butte; 26b, uncertain; 27, Steele Butte; 29b, Cedar Creek; 29c, d, and e, Pistol Ridge. Gilbert's bearings based on magnetic north.

{p. 72}   The NE spur (8) [*Horseshoe Ridge laccolith*] of Ellen Peak is slatey near the saddle and on the slopes A and B, but a well defined dike rises at C and runs down the slope to the left (NW). The mass 8 and both its spurs show chiefly trap on the surface but there is a hint that it is largely talus. To the right of 12a and then southward to 16 are a series of grass covered ridges that contain little or no trap. This soft belt is probably continuous with the great open basin (11) before me [*Sawmill Basin*] which would not have been carved so smoothly in trap.

In the narrows of 11 [*Bull Creek*] there seems to be exhibited an arch which is the continuation of the great fold of the east flank.

I suspect that the NE butte contains masses of altered massive sandstone.

The crest of 19 [*North Summit Ridge*] is slaty but thence to 20 is chiefly trap. The spur W from 19 is slaty except a few points and its extremity, whatever rock shows along the crest of a ridge, is apt to stream down both sides and ordinarily only the composition of the crest can be told.

{p. 78}   Returning I find the base of 19 is a trap (a) different from Sta. 1 and the top is a huge mass of slate dip 60° to the north. Then another trap (b) and a thin (c) which extends all the way to Stranger [*?*]. Following this a few rods beyond the Stranger I find it overlying a slate mass with a dip (45°) to the NE. Under this is still a different trap (d) which continues the ridge (with slight interruption) by variety *E* to the summit divide where are slate and sandstone with SW dip at a high angle. Summit divide at 4.50 = 20.10 = 11,800.

Thursday, September 14. From Summit Camp to the foot of Bowl Valley (camp 4).

Point 12a is about 50 or 100 feet lower than Summit Camp, i.e., it projects below there.

We find the sides of the cañon covered with trap and some slate. *Wind.* In Camp Cache, which is at the foot of the

{p. 80}   mountain slope on the west side we had a gentle breeze *up* on the p.m. of September 10 and the a.m. of September 11 and *down* the intervening night.

Gomas Camp is in the open cañon of the Cache [*Dugout*] Creek and there the wind was down on the night of the 11th.

Summit camp was occupied from 10 a.m. September 12 till the morning of 14th and the wind was down (W) all the time although part of the p.m. of the 13th the wind was E on the mountain top. Summit camp was in a narrow V-cañon near the divide and east of the crest.

Right on the line of the soft belt noted yesterday we came upon a great mass of unaltered shale with a dip to the NE. We contour it.

Camp 4   11.30 = 22.18 = 9100.

Divide (22.07) bet the Bowl and the next cañon at the south.

Climbing the NE Butte we leave our horses at 21.92 = 9450 = 12.45.

Wind. Descending the cañon today we had the wind in our faces (E) and

{p. 82}   at the same time the flag indicated a west wind.

Summit of NE Butte 2.15 = 21.00 = 10,600.

North East Butte (Bull Mountain), the summit of which is Station 2 and point 14, is a massive eruption of trap of the most massive granitic habit.

I see no rising of the strata to this Butte. Bowl Creek washes its immediate NW base [*and*] bares strata seemingly structureless with reference to the Butte. The further cliff of the stream [*looking eastward across Granite Creek*] shows a mono fold of Belted [*Morrison Formation*] dipping from the main mass of Ellen or NE. The mono *a* (Fig. 6.17) is at the foot of the mountain. The slope *b* is about the same as the grade of the creek (3° is a mere guess). The structure *a-b* can be traced halfway to the Hilloid Butte. The hill this side the Hilloid shows the lower Blue Cliff (A, p. 45) [*?*] (or else the Belted again) and the shales of the same can

Figure 6.C. Oblique air photo of west side of Mt. Ellen. Gilbert's Cache Camp was by Dugout Creek, the prominent creek discharging onto the pediment at the left center. South of the creek is the Dugout Creek laccolith and the overlying Sarvis Ridge laccolith (Gilbert's Newberry laccolite). North of Dugout Creek are the Cedar Creek and overlying Pistol Ridge laccoliths (Gilbert's Giekie and Shoulder laccolites). The prominent peak left of center is Ellen Peak. The skyline south of it is the North Summit Ridge, which ends at Bull Creek Pass on the right skyline. Right (south) of the pass is the South Summit Ridge largely formed by the shatter zone along the west side of the Mt. Ellen stock. At far left is Table Mountain (Gilbert's Hilloid Butte in his notebook and Marvine laccolite in his report); it is a bysmalith with Morrison Formation cap faulted upward around the distal (left) edge of the intrusion, which has turned-up Tununk Shale around its base. (Photograph by Fairchild Aerial Surveys.)

Figure 6.17. Cross section eastward from Gilbert's Northeast Butte (= Bull Mountain) as he inferred the structure looking down from top of the butte. His cross section is amazingly prophetic. The intrusion (c) is faulted upward and the boundary between (b) and (c) is a narrow fault zone with beds of the (b) block dragged upward against the fault at the edge of the intrusion. It is the faulted contact at the edge of the bysmalith.

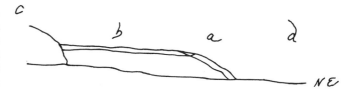

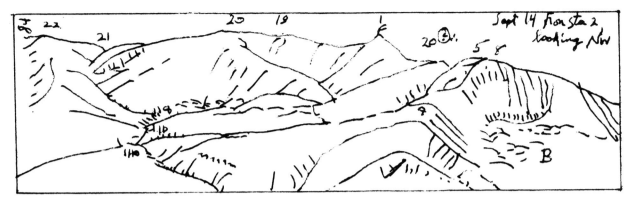

Figure 6.18. "September 14 from Sta. 2 looking NW." The view is south of west. 5, 7, and 8 are along Horseshoe Ridge. This is a laccolith with well-exposed floor at the base of the cliff where the porphyry rests on Ferron Sandstone. Area at B is old (Pleistocene) landslide. 1 is Ellen Peak. 19-20 is North Summit Ridge; 21 is Bull Creek Pass; 22 is South Summit Ridge; 13b and 12c are Wickiup Ridge; and 11a, b, and c are along Bull Creek.

be seen at *c* and *d* though in each case they are masked by foreign debris. On the plain the Belted reappears at a distance of some miles.

The peculiar crest 8-9 [*Horseshoe Ridge, Fig. 6.13*] seems to be from dike work

{p. 88}   and the depressed block is a puzzle. In detail it shows some slides. It has almost no drainage system and is not cañoned. For the present I give it up. The cañons in the face of 8 [*high part of Horseshoe Ridge*] are very peculiar. Perhaps an oblique dike has been undermined from below?

I see now the explanation of the two mouths of Bowl Cr. It is the delta principle. The delta portion being the debris cone at the mouth of the cañon and cliffs and giving distinctness to the lower benches. The main stream is now at the extreme left edge of its cone and its abandoned beds are dry washes heading on the cone and only supplied with cañon through the Belted and lower rocks. Slightly changing my perch so I can see the E base of the Butte I can see that just here there is an outbowing of the basal mono (Fig. 6.20). Succeeded southward by a reentrant curve and then by a great convex sweep. The inconsequent drainage has cut through the whole Belted into the gray sand which is here red as the Vermilion. For structural

Figure 6.20. Sketch map illustrating out bowing of the monocline at the base of Mt. Ellen to circle the northeast side of Bull Mountain (shown here at "Butte").

{p. 90}   beauty it is another Red Arena.

The angular sand derived from the trap is thrown into ripples by the wind and these are the conduits of the rain on many slopes. These give some patches of little potatoe hills (as it were) in double rows that seem to be cross rippled. Wind. In camp 4 the wind blew up till sunset and then down all night.

{p. 92}   **Friday, September 15.** Camp 4 at foot of Bowl Valley. Climbed Sta. 2 and was driven off in an hour by rain. We were visited by Graves on his way to Mt. Ellen.
Saturday, September 16. Camp 4 still held.

The storm of yesterday ceased in the night. The wind was variable but chiefly westerly. This morning it is W or down and the mist on the mts dries rapidly. The arch of rocks below camp is shown to be anticlinal by the creek section and south of the creek the dip is south. The capstone is a conglomerate that does not seem to be Belted. I think I see dikes in the cañon.

Climbing after breakfast to the top of the arch south I find that the conglomerate is a sandstone and holds about the same height N and S (Fig. 6.22).

{p. 96}   Sta. 2 (aner. 3)   9.50 = 20.875 = 10,770
                                 12.20 = 20.85
                                   3.15 = 20.75

On top of the sandstone (p. 48) [*?*] is a bed of *"Gryphaea"* several feet thick and I find an ammonite cast not in place.

The spur running from here toward 12b appears to be a dike and the anticlinal appears beyond it to the south where a cañon cuts it down to the G.M. On the other side of the cañon there seems to be only sed but on this side is either a dike or a sheet below part of the Vagabond rock.

The last word about the sandstone p. 48–50 [*?*] is that it is sandstone at top and bottom but carries a heavy bed of conglomerate in the middle. The shales immediately above the trap are exceedingly hard and withstand the weather but there is no trace

Sept 14, N. from Sta 2

Figure 6.19. "September 14. N from Sta. 2 (= Bull Mountain)." The view is about northwest. At left is Table Mountain, Gilbert's Sta. 7; see also Figure 6.12. A marks Birch Creek. The numbered points at the right are along the escarpment of Ferron Sandstone, generally referred to by Gilbert as the Lower Sand.

Sierra Abajo from Sta 2

Figure 6.21. "Sierra Abajo from Sta. 2."

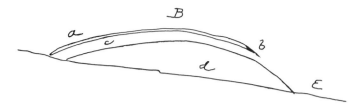

Figure 6.22. Section along Bull Creek showing the arch there. "*a* and *c* are shales. *d* is trap solid to the creek. The section from the crest of *b* at B to the creek below is 300 ft."

"The proportions of the sketch are good except that the dip is a little too great at *a*."

of foliation nor implication. The shales above the sandstone are affected but in a less degree.

Wind. Down the cañon all day. Clouds about the mountain top. Think the flag blew away last night. Trap from point 9 comes down to camp. It possibly overlies

{p. 104}   but probably traverses the shale above *"Gryphaea"*.
Sunday, September 17.
18.10, 45° = 22.710 - 7.30 a.m.

$$\underline{13.844} \quad [?]$$
$$8.866$$

2 - 22.53 =   9680
3 - 22.02 =   9310
    From camp 4 to camp 5
    After crossing a low divide and a shoulder on the slope of 2a [*12a?*] we came to a creek [*Granite Creek*] where it plunges into a cañon [*Granite Hole*]. 21.85 = 9540 = 9.15.
    The divide west of 15 at 11 am = 21.00 = 10,600 [*Pearle Flat*]
Camp 5 = 21.98 = 9580 = 12.40
    Close to camp I find *"Gryphaea"* in place. 100 ft above it is a sheet of trap that shows a slight dip to the west.
Monday, September 18. Camp 5 = 8.45 = 21.97 = 9590.
    Climbed sta. 3-pt. 15. There are some white crags a little below this on 12a that I suspect are sandstone

{p. 106}   but they are small in extent compared with the trap. [*Bull Mountain, well shown in Figure 6D, is a bysmalith. The nearly horizontal beds of the Morrison formation that are well exposed in the ledges around its base do not extend under the porphyry; they are turned up vertically against the upward faulted edge of the intrusion. This structure is expressed physiographically—the resistant diorite porphyry would not have been eroded back off the top of the more easily eroded ledges of shale and sandstone. See Figures 6.18 and 6.27.*]
    From this point there seems to be no question that the strata run into NE Butte without uplift by its trap. I make no synclinal south of the Butte (such as is north), the two arches are connected by a straight line (Fig. 6.27).

This spur makes a dike and there are branches to it but the strata which appear at its eastern spur run independently and dip gently toward the main mass of Ellen or away from the eastern arch. East of the spur a narrow dike with E-W trend shows in the heart of the arch.

All along the western limit of the arch, conforming sheets of trap appear dipping toward the mountain, but only on Bowl Creek have I yet seen a core to the eastern arch [*western arch?*].
    Station 3 at 1.30 = 20.75 = 10,950.

{p. 112}

Cash on hand August 15		$207.15
Aug. 15	Overalls	2.00
	Hotel Ogden	1.25
	lunch, Cache Valley	.30
Aug. 16	Horse hire (Jno Boice v)	7.00
Aug. 18	Fare. Franklin to Ogden	4.00
	Hotel Franklin	2.00
	Lunches Ogden	.90
	Papers	.25
Aug. 19	boots	7.00
	deposited in Deseret Nat. Bk.	125.00
	' stamps' 10 'spurs' .75	.85
	'semper cerati' .10 books 1.90	2.00
	Hotel Salt Lake	6.00
Aug. 20	fare to York	4.00
	paper	.10
	billiards	.50
	Dinner at York	.50
Aug. 27	lodging at Chicken Creek	1.50
	milk at Fayette	.25
Aug. 27	Washing Gunnison	1.50
Aug. 29	'Sermm' a tract	.70
	washing	.50
	'AHT' Cr	300.00
Oct. 8	Averitt Dr	5.00
Oct. 8	Farnsworth Dr	5.00

{p. 110}
Letters

Aug. 15	No. 16
Aug. 14	GSG (insurance)
Aug. 17	No. 17
Aug. 19	No. 19, Mrs. Smith
Aug. 21	No. 20 (card)
Aug. 23	No. 21 (Mrs. Brown)
Aug. 22	No. 8 recd
Aug. 24	Nos. 7, 9, 10 recd
Aug. 25	No. 11 recd
Aug. 25	No. 22
Aug. 26	Emmas
Aug. 27	No. 23
Aug. 29	'frm' Emma
Aug. 30	No. 24 and Emma

130

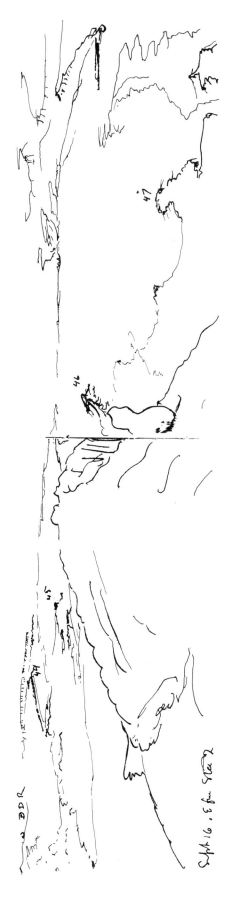

Figure 6.23. "September 15. SSE [*actually SSW*] from Sta. 2." 12a and b are along Wickiup Ridge. The point marked 19 probably should be 24; see Figure 6.15. 17 is Barton Peak. 15 is along Granite Ridge. Points numbered at left are in the Maze Arch.

Figure 6.24. "September 16. E from Sta. 2."

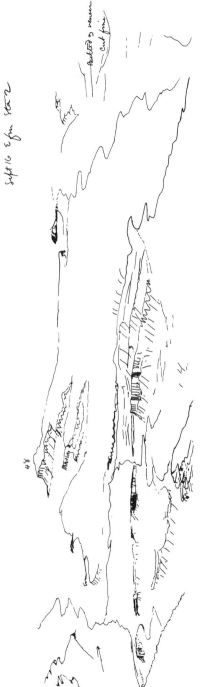

Sept 16 E fm Sta 2

48

Figure 6.25. View east from Station 2, top of Bull Mountain. Due east from the station is along the fold between Gilbert's two pages. Point 48 is Goatwater Point. Granite Creek enters the scene at the lower right and flows north in a strike valley along the hogbacks (Morrison Formation dipping east away from the faulted east wall of the Bull Mountain bysmalith). A little north of the location of Granite Ranch (position indicated by coordinates GR) it is joined by a tributary from the west and turns east and then northeast to join the Dirty Devil River (out of sight). The wash draining east across the middle of the right half of the drawing is Poison Spring Wash. During the early Pleistocene all this area drained eastward via Poison Spring Wash, but that ancestral drainage was captured and turned northeastward by Granite Creek. The canyon by which Poison Spring Wash joined the Dirty Devil is the largest western tributary canyon of the Dirty Devil, but now it receives practically no drainage from the Henry Mountains.

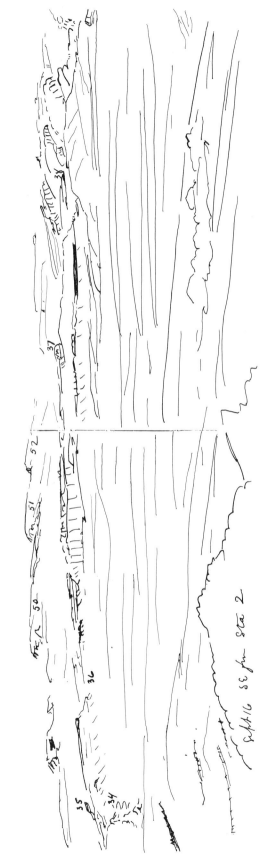

Sept 16 SE fm Sta 2

Figure 6.26. "September 16. SE from Sta. 2." Gilbert's points 35 to 38 are along the scarp of the Morrison Formation (Gilbert's Belted Formation) encircling the Maze Arch. Points 51 and 52 are along the rim of the Morrison Formation where it begins its rise eastward from uplift at Mt. Ellen.

Figure 6.D. Oblique air photo of the east side of Mt. Ellen. Ellen Peak, Gilbert's Sta. 1, is the sharp conical peak right (north) of center. South of it is the North Summit Ridge, Bull Creek Pass, and South Summit Ridge, the latter formed by the shatter zone around the Mt. Ellen stock in the Bromide Basin (most southerly canyon). Below and to the right of Ellen Peak is a curved ridge, the Horseshoe Ridge laccolith. Gilbert's points 8 and 9 in Figure 6.13 were on this ridge. Below it and slightly to the left can be seen the arch of the Bull Creek laccolith (see Gilbert's drawing, Fig. 6.12). Still nearer the observer is Gilbert's NE Butte (his Sta. 2 and point 14). In his report he called this the Jukes laccolite. It is Bull Mountain and formed by a bysmalith. Horizontal ledges of the Morrison Formation around the base of the butte seem to pass below the diorite porphyry but in fact they are dragged upward and cut off discordantly by the faulted side of the bysmalith. A sill can be seen extending NE into the ledges. The bysmalith is at the distal end of an anticline extending under Wickiup Ridge back toward the stock. A second bysmalith, Table Mountain (Gilbert's Hilloid Butte and Marvine laccolith) is at the northwest corner of Mt. Ellen beyond, and on line with, the north side of Bull Mountain. In the distance above Ellen Peak is the Waterpocket Fold and on the skyline the Aquarius Plateau. (Photograph by Fairchild Aerial Surveys.)

Aug. 30   'frm' GSG and No. 12
Sept. 1   No. 24½ and Emma
Sept. 8   No. 25
Oct. 8   No. 26, Emma, GSG, Mrs. Graves,
               Powell

Gun No. 3539

washing at Gunnis

1 night shirt

5 pr sox

7 hanx

6 'Bihd'

2 under

——————

voucher books

tape line

shovel

axe

pick

rope

hobbles

paper

ammunition

ore sacks

Rations drawn August 30. Supposed to be 160 rations

200	lbs flour in 4 sacks
53	lbs pork
90	lbs jerk 2 sacks
10	lbs rice
22	lbs coffee
1½	lbs tea
10	lbs brown sugar
30	lbs white sugar
8	lbs soap
10	lbs salt
18	cans yeast powder (½ lb Prices)
4	boxes matches (1,000?)
6	bottles pickles
10	lbs apples
15	lbs prunes

Figure 6.27. View of Bull Mountain (Gilbert's Sta. 2 or NE Butte) from the southwest. "From this point there seems to be no question that the strata run into NE Butte without uplift by its trap. I make no synclinal south of the butte (such as is north), the two arches are connected by a straight line." Compare Figure 6.17. As Gilbert noted the roof is not arched; it was lifted by faulting. Bull Mountain is a bysmalith.

10	lbs currants
1	lb 'salnalue'
2	papers pepper
1	bottle pepper
15	tobacco plugs
1	ham
8	cans milk
30	lbs beans
6	lbs dried corn

{p. 108}

60 flour	matches
25 pork	pickles
25 jerk	4 apples
2 rice	5 prunes
8 coffee	4 currants
1 tea	'salnalue'
4 brown sugar	pepper
12 white sugar	mustard
3 soap	tobacco
4 salt	ham
4 lbs yeast	milk
	12 beans
	corn

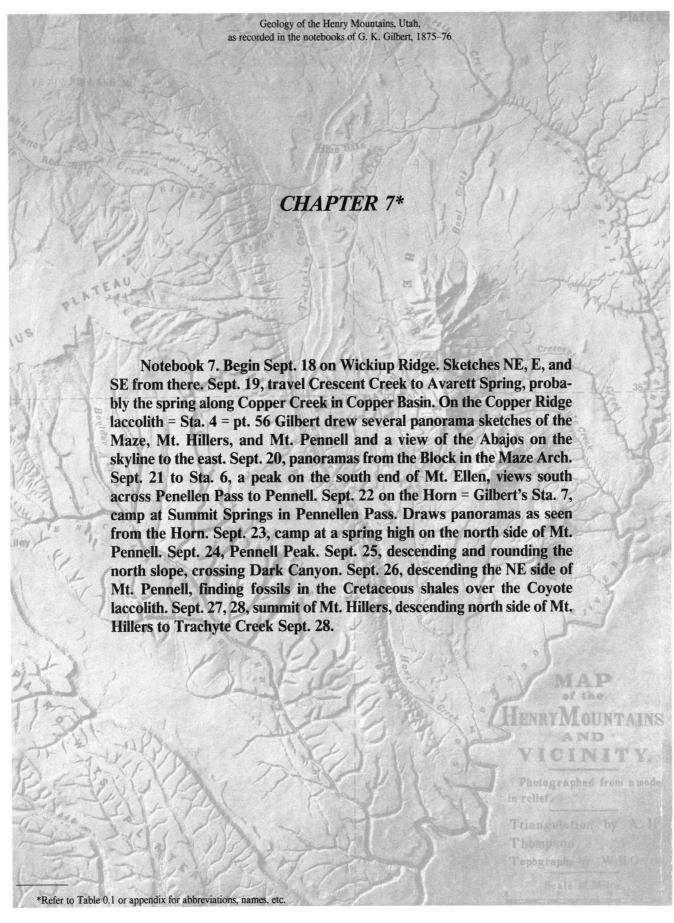

CHAPTER 7*

Notebook 7. Begin Sept. 18 on Wickiup Ridge. Sketches NE, E, and SE from there. Sept. 19, travel Crescent Creek to Avarett Spring, probably the spring along Copper Creek in Copper Basin. On the Copper Ridge laccolith = Sta. 4 = pt. 56 Gilbert drew several panorama sketches of the Maze, Mt. Hillers, and Mt. Pennell and a view of the Abajos on the skyline to the east. Sept. 20, panoramas from the Block in the Maze Arch. Sept. 21 to Sta. 6, a peak on the south end of Mt. Ellen, views south across Penellen Pass to Pennell. Sept. 22 on the Horn = Gilbert's Sta. 7, camp at Summit Springs in Pennellen Pass. Draws panoramas as seen from the Horn. Sept. 23, camp at a spring high on the north side of Mt. Pennell. Sept. 24, Pennell Peak. Sept. 25, descending and rounding the north slope, crossing Dark Canyon. Sept. 26, descending the NE side of Mt. Pennell, finding fossils in the Cretaceous shales over the Coyote laccolith. Sept. 27, 28, summit of Mt. Hillers, descending north side of Mt. Hillers to Trachyte Creek Sept. 28.

*Refer to Table 0.1 or appendix for abbreviations, names, etc.

135

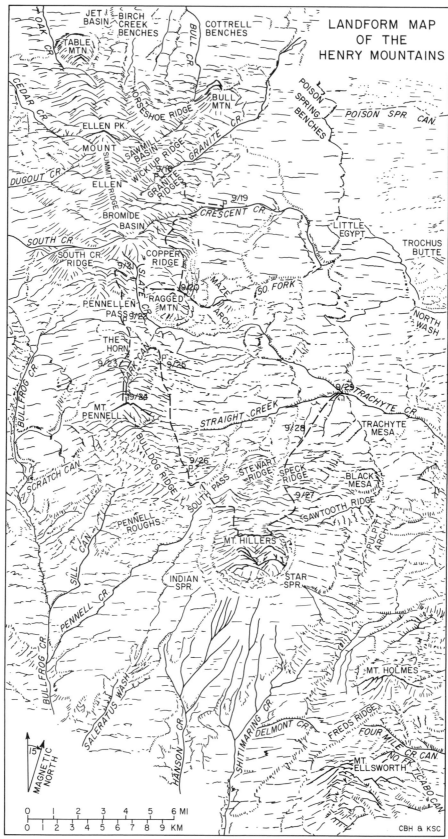

Figure 7.A. Landform map showing Gilbert's route September 18 to 29, 1876.

Figure 7.B. Oblique air photo looking north along the east side of Mt. Ellen. The dikelike mass in low center foreground is top of Ragged Mountain (Gilbert's Scrope Butte), the bysmalith at the southeast corner of the Mt. Ellen cluster of intrusions. Just beyond and right of center is the Copper Ridge laccolith (Gilbert's Sta. 4; see Figs. 7.4 to 7.10). Bull Mountain (= Gilbert's NE Butte, Juke Butte in his report) is at the right. The Ellen stock is in the Bromide Basin, in the horseshoe ridge opening east just below the 3 high peaks on the skyline. (Photograph by Fairchild Aerial Surveys.)

**G.K. GILBERT, POWELL SURVEY, BOX 806,
　　WASHINGTON, D.C.
NOTEBOOK NO. 7, 1876
OPENED ON STA. 3, SEPTEMBER 18**

{p. 4}　　**Monday, September 18,** cont. from Bk 6.

Descending the spur eastward we come to slate (a) at 20.78 = 10,900. It is level bedded and interrupted by one cross dike. At 20.85 = 10,800 a *"Gryphaea"* bed.

Talus conceals to 21.21 = 10,530 where is a sheet of trap (spec. b) 30–40 ft thick. This has a strong dip to the N and is very likely continuous (though not traceable) with A, p. 1 (Fig. 7.1).

There certainly are two sheets as sketched, lying nearly level and on the S flank of the W spur of the NE Butte. They interrupt the Vagabond Series.

Under the sheet (at 10,330) is a series of flesh colored conglomerate and red and gray shales, clearly the Belted, consistent with the Vagabond rocks below. These are very much higher than below camp five (Fig. 7.2) or even under Sta. 2. These are interrupted by sheets a little way to the west and the spur we descend seems to

{p. 5}　　be part of a massive injection that has tilted the strata to the north in the east end of Sta. 3 [*Wickiup Ridge*]. The Belted rocks are little altered in the main. In a butte a little to the south there is a sheet (?) dipping SW and apparently the reverse of the dip I have crossed. There is probably a local bulge structure greatly masked by talus and inconsequent erosion.

At 22.02 a white to flesh conglomerate about level and changed to quartzite and containing a fresh bone. The rock next

Figure 7.1. "September 18. NE from Sta. 3." View to Bull Mtn. (Sta. 2) from Wickiup Ridge.

Figure 7.2. "September 18. S from Sta. 3." The coulee passing camp 5 is Crescent Creek. Pt. 56 is Copper Ridge; 39 is The Block in the Maze Arch.

to this is less altered. I get a view later of its mass and see that the dip is about the normal one to the west.

Tuesday, September 19, 1876. From Camp Pleasant (No. 5) [*On Crescent Creek benches; see Fig. 7.2*] to Camp 6 at Avarett Spr.

Station 4 = pt. 56 [*Copper Ridge laccolith*]

In climbing I touched *"Gryphaea"* twice close to camp. Either there are two horizons for the species or there have been slips here. It is noteworthy that no fault has yet been

{p. 6}   seen in the Henry Mountains. Higher I found slate, a little sand, and much loose trap. None in place however till I came upon the shelf that ends here. Under the face of sta. 4 are both sheets and dikes that defy classification. The lowest sheet must be 100 feet thick and is below the Belted Conglom. The descent is very steep and beneath is a Red Maze wherein the J.T.

[*Jura-Trias*] is brought up by the complicated eastern fold and deeply scored by erosion. I see no dike or sheet in the Maze.

Sta 4   9.15 – 21.05 = 10.520
           11.30 – 21.06 =

I think this sta. is over a boss dike from which sheets run E and west.

Avarett Spr. 5.30 p.m., 21.26 = 10.270

{p. 14}   **Wednesday, September 20.** Avarett Spring (Camp 6) is just north (½ mile or less) of point 56 [*probably the spring in Copper Basin*].

Camp 6. 7 a.m. = 21.30 = 10,210
Sta 4 7.40   = 21.11

Left horses a few feet below *"Gryphaea."* 8.20 = 21.78 = 9620

We reach conglomerate at 21.86 = 9510. At 22.14 = 9170 we

Figure 7.3. "September 18. SW from Sta. 3." Pt. 17 is Barton Peak.

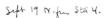

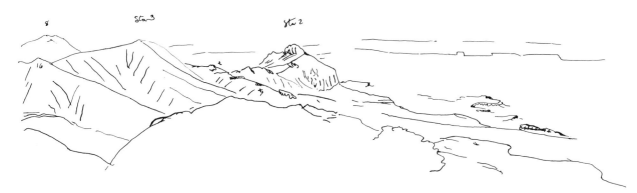

Figure 7.4. "North from Sta. 4." Sta. 2 is Bull Mtn. Sta. 3 is Wickiup Ridge.

Figure 7.5. "September 19. The Maze from Sta. 4 [*on Copper Ridge laccolith*]. 1 = Pt. 39 = The Block.

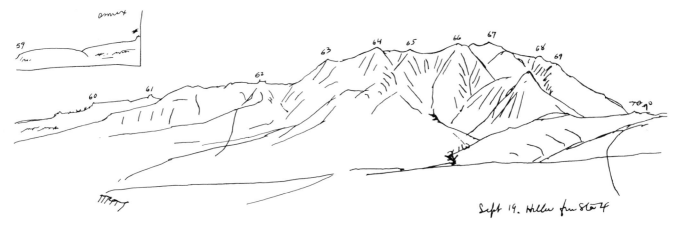

Figure 7.6. September 19. View S. to Hillers from Sta. 4 on Copper Ridge laccolith.

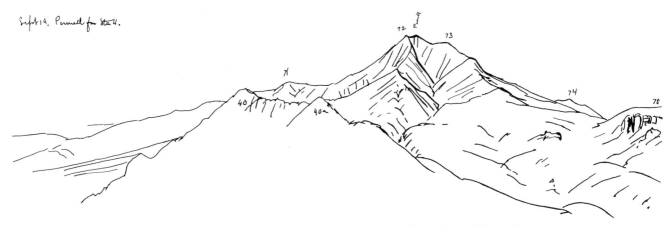

Figure 7.7. "September 19. Pennell from Sta. 4." 40 and 40a are Ragged Mountain.

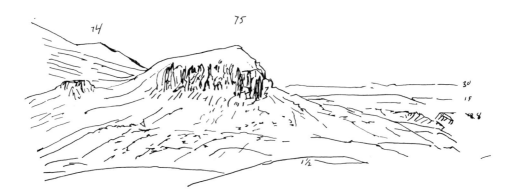

Figure 7.8. "September 19. The Pass Butte (75) from Sta. 4." (Pass Butte = the Horn.)

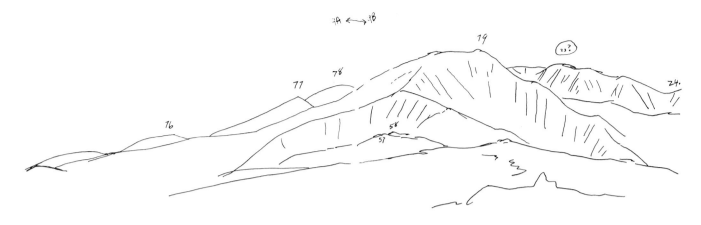

Figure 7.9. This drawing, on p. 12, is not labeled, and the editor is not sure what it shows. The numbered points and the context suggest a view N. from near Ragged Mountain.

Figure 7.10. "September 19. The country east of Henry. Sketches from Sta. 4."

seem to be below all the conglomerate and have red sandy shales.

A sheet of trap 30 ft thick comes in at 22.24 = 9040.

At 8.45 I read 22.48 = 8750 having passed 1/3 through the red sandy shale and 2/3 through a massive sandstone since leaving the trap.

From the horses we have travelled all the way cataclinal the dip being 7° by guess and our course inclining 20°. Now we rise without much change in horizon to pt. 39 = Red Head [*The Block*].

Station 5 is 50–75 ft below the top of the Head and is at the SE angle. At 9.30. 22.34 = 8930. The cliff of Belted conglom. continues all along under sta. 4 and c. (Fig. 7.14)

{p. 15}   to 40a [*Ragged Mtn.*] and then ends abruptly against a fault as I supposed (A p 16 = Fig. 7.12) being resumed at BB under 40 (p. 16). The same rock appears beyond Trachyte Creek at C (p. 16–17) (Fig. 7.12).

Toward Pennell the eastern mono of the east arch seems to

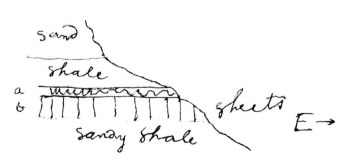

Figure 7.11. Sheets of diorite porphyry (Gilbert's trap) under the Copper Ridge laccolith at Sta. 4. "The lower one 30 feet thick and pale; the upper 10 ft and dark."

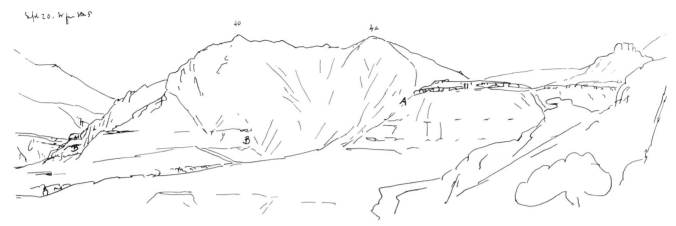

Figure 7.12. "September 20. W from Sta. 5." (Sta. 5 is The Block in the Maze Arch = pt. 39 in Fig. 7.2.) The view is to Ragged Mountain. B-B, Morrison Formation. 4a should be 40a.

Figure 7.13. "S from Sta. 5. September 20." C-C-C, upper part of Morrison Formation closing around the north end of Maze Arch.

flat out and merge with the general eastward dip of Pennell, but the arch is not lost and its crest swings to the right and goes straight for Pen. See p. 17 (Fig. 7.13).

Sta. 4 is on a lenticular sheet with others below (see p. 18) (Fig. 7.14). They must thin E as well as N and S or there would be cliff erosion. Indeed their thinning is visible. Still no trap seen in the Maze.

In returning to our horses we climb the spur A (p. 18) and look across at the sheets B.

There are two sheets, the lower one 30 feet thick and pale, the upper 10 ft and dark (Fig. 7.11).

I call the upper the newer because it was lighter in weight melted than the other cold. Similarly D is darker than E and the differences in time may be identical.

Later I collect *a* and E and find little difference. The diff in

{p. 19}   color is a matter of weathering.

We return to the horses and leave the spot at 2 p.m. reading 21.82 = 9580. *Wind* at camp 6 was very gentle but S or down the cañon. At camp 7 at summit today (S 20) it is N or down the cañon. [*Camps 6 and 7 were on opposite sides of Copper Ridge.*] **Thursday, September 21.** Camp 7.   7.5 a.m. = 22.17 = 9120. Sta. 6.   8.20 = 20.79 = 10890.

There is the same indefiniteness about the structure of Pennell that I found last year. The topography of the upper portion gives no clue to its structure [*a stock surrounded by shatter zone; Fig. 0.D*]. The strata rise toward it from east and west and the Maze fold enters it from the NE or N but they cannot be traced through. If we produce them through (Fig. 7.15) the dips from the two sides correspond nicely and the top of the upper blue cliff runs to the summit of the mountain. This may easily be the

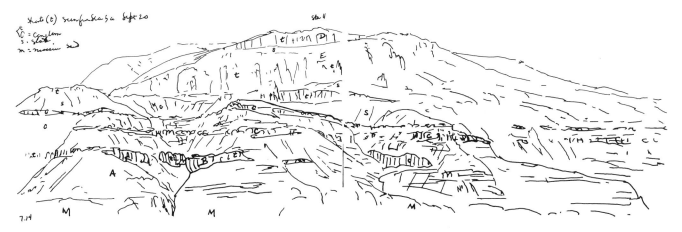

Figure 7.14. "Sheets (T) seen from Sta. 5 Sept. 20." (Sta. 4 is the Copper Ridge laccolith intruded toward the observer from the Mt. Ellen stock; it is in the Tununk Shale. Most of the associated sheets are in the Morrison Formation.)

structure with a few lenticular sheets to lift the crest of the arch and a core of dikes preserve the crests from erosion. Thus Ellen and Pennell would seem more closely

{p. 20}   united than other individuals of the group. [*The structure contour map of the mountains, Fig. 0.D, indicates that Mt. Pennell is quite independent of the Mt. Ellen intrusive masses, as independent as is Mt. Hillers.*]

The Pass Butte [*the Horn*] is a local pocket of trap (Fig. 7.16).

Sta. 6 = pt. 76 is a narrow crest due to a dike.

Descending at 21.80 = 9590 I pass the top of a sandstone that is probably the 1.b [*Lower Blue*]. At camp 7 a few minutes later 22.16 = 9140. The outcrop dips south and there is another

Figure 7.15. The three major Cretaceous sandstones projected over Mt. Pennell.

Figure 7.16. "September 21. From Sta. 6. Looking south." (Sta. 6 not positively identified but I have judged it to be the knob just under 9,600 feet altitude a mile west of south of Kimbell Turner Peak on the south rim of the Bromide Basin. In his report, Gilbert anticipated the later discovery that the laccoliths are tongue-shaped masses forming anticlines radiating from the stock.) He wrote (1877B, p. 38), "In addition to the laccoliths of the foundation and of the main body, there is a series which juts forth from the northern flank like so many dormer windows." He illustrated the point with a drawing like this of the Horn Laccolith, his Sentinel Butte.

Figure 7.17. "September 21. W from Sta. 6." (95 to 98, 101, and 119 are along the Emery, Gilbert's C, or Gate Sandstone, as are 113 to 118 in the nearest hogback marked "g" and bench to the south. 103 to 109 are rim points on Tarantula Mesa.)

Figure 7.18. "September 21, from Sta. 6." (Looking east to Ragged Mountain pt. 40, 40a.)

60 ft lower down the hill. They make ledges down which a way has to be selected. I detect a dip at A p. 21 (Fig. 7.16). It is from Pennell and from the Maze Arch.

{p. 24}   **Friday, September 22.** Camp 8 at Summit Spring.
6.50 a.m.   1810 – 50° 22.798 (13931)
                        3 – 22.07 = 9260
                 cerrostrat 80%. Wind W.1
Sta. 7 = Pass Butte [*The Horn*] = pt. 75
at 9.50 a.m. = 21.03 = 10,570.
at 2.40 p.m. = 21.00 = 10,600.

In the foreground of sketch on p. 25 (Fig. 7.19) all the sed E of A is Belted and all west is lower blue. I do not separate the *"Gryphaea."* Both dip south and the 1.b [*Lower Blue*] dips west also. The Belted does not rise to 40 [*Ragged Moutain*] but on the west rather dips toward the Butte [*Ragged Mountain or the*

Horn?]. Under the Butte at the south it seems to dip east, and the cañon reveals the G.M. below it. There are hints at bedding all across the end of Ellen and the indication is that the lower sand does not reach the summit but that the rocks there are 'gate' or even upper blue. The summit itself however is entirely noncommittal.

The block on which I stand has its counterpart in one through which the drainage from Pennell cañons. The two were not connected; the slope is open between them. [*Gilbert refers here to the Dark Canyon laccolith, which is next east of the Horn laccolith.*] In common with 40 [*Ragged Mountain*], 56 [*Copper Ridge*], and 14 [*Bull Mountain*], they were formed entirely above the Belted. There is a smaller one 2 miles to the SW (A, p. 21) (Fig. 7.16).

The lava of this butte is really different from that of Ellen. Its feldspar is pinkish and mica abounds. Think there are two layers

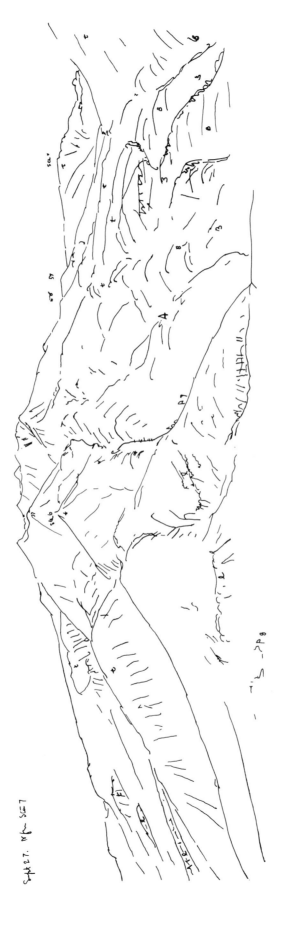

Figure 7.19. "September 22. N from Sta. 7." (This is view from the Horn north to the south end of Mt. Ellen. The creek at Camp 7 is Slate Creek; it heads against the south rim of the Bromide Basin near pt. 77 rather than breaching it as suggested in this sketch. On skyline right at Sta. 4 is the Copper Ridge laccolith extending south of east from the Mt. Ellen stock in the Bromide Basin, high ridge, center; on the skyline left, not marked, is the South Creek Ridge laccolith extending west from the stock. The ridge extending down from Sta. 6 and the ridge marked "t" between it and the South Creek Ridge also are laccoliths radiating from the stock. The butte at the right designated "t" for trap is the Ragged Mountain bysmalith breaking up through the Morrison Formation = Gilbert's Belted (B).)

Figure 7.20. "September 22. NW from Sta. 7." (This is view west from the Horn. Compare with Fig. 7.17).

Figure 7.21. "September 22. 40 from base of Sta. 7." (View is east from base of the Horn to Ragged Mountain and the ledges at the head of Slate Creek. The ledges are Morrison Formation, Gilbert's Belted; Ragged Mountain is a bysmalith that has broken upward by faulting. The porphyry does not rest on the horizontal bed but cuts upward through them along a steep faulted contact.)

{p. 28}   the upper of which has a shelly cleavage while the lower is massive. Collecting the lower I find it whiter (perhaps a difference in weathering) and without mica. I collect also the lava of a smaller dike or sheet at the east. Under the latter I find shale that seems almost or quite unaltered. This accords with the breaking off of the Sentinel in cliff change.

Saturday, September 23. Camp 8 at Summit Spring 7.30 a.m. 22.06 = 9280.

Climbing 40 = sta. 8 we leave the Belted Conglom (quite quartzitic) at 22.26 = 9020 and reach *"Gryphaea"* at 22.02 = 9320. The separation is exaggerated by the dip which rises with us. The *"Gryphaea"* is somewhat indurated.

Left horses 9 a.m. at 22.00 = 9350. At 21.84 is 'at times'

arenaceous shale in place, and I see boulder of conglomerate a little below.

At about A, p. 25 (Fig. 7.19), the change from lower blue to Belted was made by a mono fold with a throw of a few hundred feet to the west.

The gap between is partly

{p. 29}   structural and partly eroded. There is a measure of dip in that way—enough to direct nearly consequent drainage. But there is also an absence of the 'exposure' induration which holds the shales of Pennell and Ellen.

Summit 40 [*Ragged Mountain*] = sta. 8, 10.10 a.m. 21.09 = 10,490.

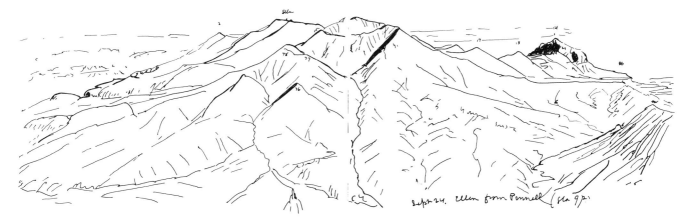

Figure 7.22. "September 24. Ellen from Pennell (Sta. 9)." Compare Figure 7.19.

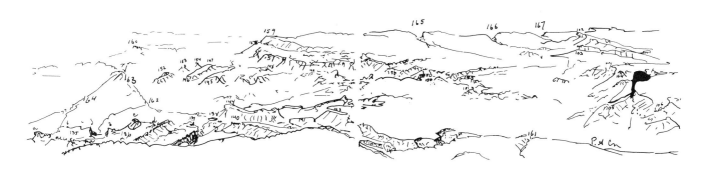

Figure 7.23. "The Gate Cliff south from Pennell. September 24, 1876." (Tarantula Mesa is at the right. South of it is the broad bench formed by the Emery Sandstone = Gilbert's Gate Sandstone. P. A. Cn = Pine Alcove Canyon = modern Bullfrog Creek.)

There is a very even bench around the base of the Sentinel [*The Horn*] at A A (Fig. 7.16) that seems to indicate bedding though no bed is visible. It shows a rise from the saddle toward Pennell just as toward Ellen.

Topog note. This point, camp 8, and pt. 116 are about in line. Will fix camp 8.

Horses again 12.50, al. 97 = 9390 [*upper 1,000 ft of Ragged Mountain on foot*].

Camp 9 is on the N base of Pennell W of the cañon. At 7.40 p.m., 21.10 = 10,470.

{p. 30}　　**Sunday, September 24.** Camp 9.　　6.15 21.10 (?) = 10,470.

Slate and trap. The slate on edge trending N. Thin spur 7 am, 20.02 = 11,910. We leave horses on saddle close to summit. 7.35 - 19.65 = 12,400.

Pennell = sta. 9 at 7.45, 19.49 = 12,630. 1.15 p.m. = 19.44. 2.50 = 19.42.

The vertical (or nearly vertical) slate on the spur of Pennell is without precedent although 60° dip was noted on Ellen. It is just possible the north end of Pennell is due to a fracture but it is improbable. There is little trap with the slate and the summit as usual is of trap. The other spurs look similar; slaty and trappy. [*Gilbert is looking at the shatter zone on the N side of the Pennell stock.*]

There certainly is an old erosion base marked by the summits of the Gray Cliff rock from the vicinity of Sorenson's pocket southward to the formation islands [*Circle Cliffs*].

At the south toward Navajo Mt. there is a broad low flexure with S. throw (or else a synclinal with SW trend) which prolongs the cañon of Pine Alcove through the Belted

{p. 33}　　(from p. 30) and even diminishes its depth in front.

The south spur of Pennell shows a tendency to N–S structure, but I cannot tell whether I see strata or sheets or dikes.

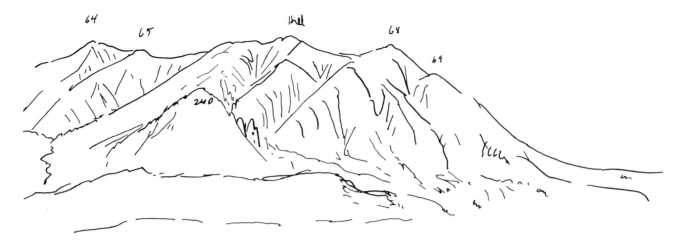

Figure 7.24. "September 24. From Pennell." (View of Mt. Hillers. Pt. 240 is Bulldog Peak. The numbered peaks on the summit are along the north edge of the Mt. Hillers stock. From its south base to this summit ridge the stock and its shatter zone cut discordantly from Permian to Cretaceous formations, more than a mile of discordance.)

There are some good type badlands on Divide Creek and on the other side of Pine Alcove below the Gate Cliff. In the former case the only cap

{p. 34}   to be removed is a veneering of trap, the sandstone (2nd B.G.) having been planed off when the surface was veneered.

There is much consequent drainage about these mts. The drainage of the east flank of sta. 2 [*Bull Mountain*] is consequent to that bulge but is inconsequent to the one through which it cuts [*Bull Creek Arch*]. Hence the former is the newer. (???) [*Gilbert's queries; he didn't believe this either.*]

There are no definite dips on this side of Hillers—seen from here.

Descending the Mt I collect at several points the trap of the spur.

Monday, September 25, 1876. Camp 9 is close to the N cañon of Pennell. At 7.45 a.m. 21.22 = 10,310.

When we cross the cañon read 21.51 = 9970.

A mile from the cañon we came upon sandstone at 21.36 = 10,140 and we meet it several times thereafter.

At 21.90 I am on sandstone which I sketch the following (p. 34; Fig. 7.25) on a spur leading direct from 72 [*Pennell Peak*]. There is a great mass *ab* with obscure horizontal bedding but no intermixed sedimentaries

{p. 35}   and some overlying sheets too even to be anything but conforming. I think the sandstone on which I stand belongs in the gap *d,* and the letter *d* shows about my height in the section. This dip strikes toward pt. 134 [*? 134 not located*] and can be traced some distance. It forms part of the east flank of the east fold.

Camp 10 on the joint fork of Trachyte Creek, 11.45 a.m. 22.41 = 7625.

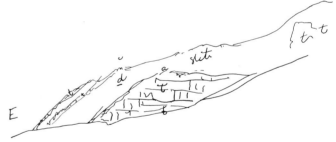

Figure 7.25. "September 25. Sketch referred to on p. 34."

{p. 36}   **Tuesday, September 26.**

Camp ten is in Cret. Shale unaltered. I saw on a loose piece a fragment of *"Inoceramus"* 17 inches in diameter. I think the dip is from Hillers.

Camp 10: 7 am - 23.41 = 7650. Sta 10. 9 a.m. = 21.86 = 9510.

The ascent to sta 10 is over a series of slidden and fractured masses of at first sandstone and then trap, and a good deal of the country to the E and S of us has a slidden look.

The shale at camp 10 is the Blue Gate and near the base. The sandstone fragments I have climbed over are of the Blue Gate. The line of cliffs 2 miles east of camp is lower Blue. The Gate sand makes a ridge in the Hill-Penn Pass with a dip half to the east and half from Hill.

149

Figure 7.26. "The Maze from Sta. 10." (19 probably is Barton Peak formed by the shatter zone on the east side of the Mt. Ellen stock; 40 is Ragged Mountain, a bysmalith that is faulted upward across the ledge of Morrison Formation around the base of the hill. 56 is the Copper Ridge laccolith. 14 is Bull Mountain.)

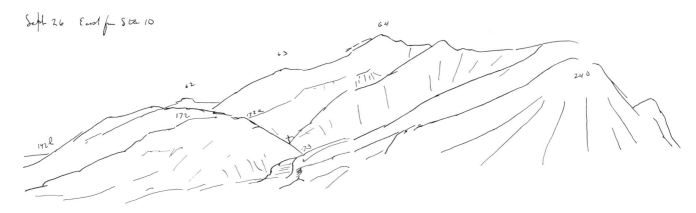

Figure 7.27. "September 26. East from Sta. 10." (View is along the NE side of Mt. Hillers. Pt. 240 = Bulldog Peak of U.S.G.S. Prof. Paper 228.)

Sta. 11 at 2.15 = 2192 = 9440.

There is sandstone 200 ft below Sta. 10 and it dips S and E making a synclinal with those of sta. 11. The structural saddle is about a mile north of the pass and is probably somewhat west of it. That is to be seen from Hillers.

I think both these are the lower but that too must be seen from a point further south. They may not be the (Figs. 7.26, 7.27, and 7.28a and b)

{p. 40}   same and in that case this is the lower. The dips fluctuate but do not exceed 10° anywhere.

There is an unmistakable terrace (see p. 39) on the W flank of Hillers but I see no seds in it. There are probably sheets with a gentle dip.

240 [*Bulldog Peak, Fig. 7.27*] is a bulge from which all the seds are eroded—'in figs'.

Hill-Penn Pass at 2.45 = 22.37 =8900.

Camp 11 at 3.35 p.m., 22.26 = 9020

Sta. 11 like sta. 10 is on trap 'slope' over sandstone.

Looking at Hillers from camp 11 I see nothing that I can call 'any but' trap.

Wednesday, September 27.

6.15 a.m. Ann. 3 22.17 = 9140

            2 22.63 = 8560   (18.10 AE) 'Camp and sing' at camp 11.

      Saddle (172a)   7.30 = 21.67 = 9700

On a level with the top of 172 I read 21.52 = 9950

At 7.50 = 21.15 = 10,440 I am on the sharp angle of the spur I climb. Higher than the saddle of 240 but still lower than 240 (Fig. 7.27).

{p. 42}   Sta 12 = pt 64 = East summit of Hillers.

 9 a.m.   = 20.18 = 11,700
10.45   = 20.18
12.40   = 20.12
 2.20   = 20.09
 3.40   = 20.09

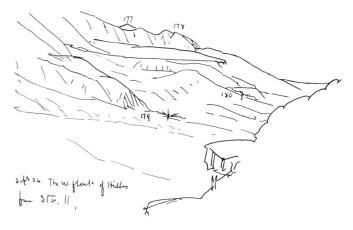

Figure 7.28A. "September 26. The W flank of Hillers from Sta. 11."

Figure 7.28B. "September 26. N flank of Hillers from Sta. 11."

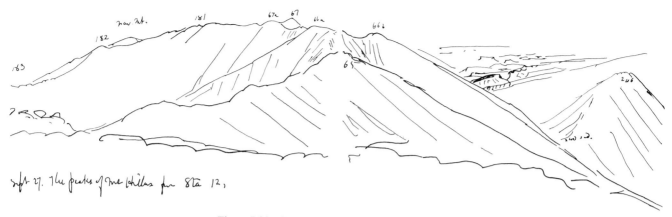

Figure 7.29. "The peaks of Mt. Hillers from Sta. 12."

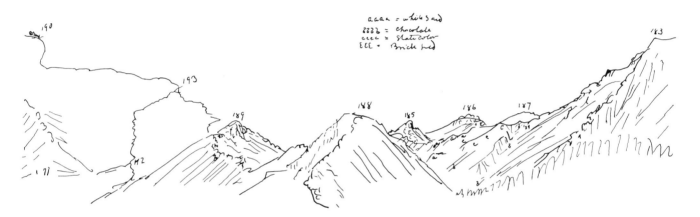

Figure 7.30. Unlabeled sketch of view south from Sta. 12, easternmost of the high peaks on Mt. Hillers.

Here for the first time I get sight of a spur on the west flank of Pennell which consists of an uplifted Cretaceous sandrock (223, p. 47) (Fig. 7.33). It can be well studied from 68 or 134.

Near camp 11 there is a spur capped by a trap sheet that shows a decided dip conforming with the sandstone of the Pass ridge; i.e., it is toward Pennell. I see no sympathetic symptoms in other topography of the vicinity.

The sandstones under sta. 10 and 11 are identical and the synclinal is close to the pass at A p. 49 (Fig. 7.35). There can be no question that is the Gate sandstone. The shale under it is continuously visible till it overlies the lower Blue down Trachyte Creek, while the shale below that is traceable by the eye to the Belted (Figs. 7.29, 7.30, 7.31, 7.32, 7.33, 7.34, 7.35, and 7.C).

{p. 50}   **Thursday, September 28.** Camp 11 (Appendix B). 7 a.m. = 22.18 = 9100
pt. 65 at 8.40 = 20.28 = 11,570

The 65 (Fig. 7.29) spur has trap only on its crest but the ridge connecting it with 66a (Fig. 7.29) shows (beside) trap much green and purple shale. 66a shows conglomerate of a purplish tinge at base. In fine the rocks are of the horizon of Belted conglomerate. I think the shales are in place. They are a little altered—just enough not to make mud. Think they are about level.

Leave horses at 9.15 at 20.23 = 11,130. At 20.15 = 11,230 we are on gray shale in place dipping N at 15°. Above this is a cap of trap with the same dip and forming the N part of the summit of 66a.

66a at 9.35 = 20.02 = 11,920.

The highest crest is of slate dipping NNE 20°.

Station 13 = Pt 67 = summit of Mt. Hillers.

10.5   am = 19.95 = 12,010
11.15 am = 19.93
1.   pm = 20.00
3.10 pm = 19.98
4.   pm = 20.01

Figure 7.31. "September 27. From Sta. 12." This view is southward along the east flank of the Henry Mountains syncline, but I am unsure which benches are recorded in the sketch.

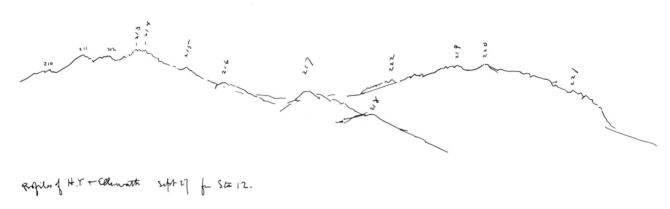

Figure 7.32. "Profiles of H.V. [*Mt. Holmes*] and Ellsworth. September 27 from Sta. 12."

Figure 7.33. "September 27, from Sta. 12." (The view is NW to Mt. Pennell. 240 is Bulldog Peak; 223 is the big sill dipping off the SW flank of the Mt. Pennell stock. Extending north from the Pennell stock is the ridge formed by the Dark Canyon laccolith and below it to the north is the Horn.)

Figure 7.34. "September 27. The Maze fold from Sta. 12." This sketch was reproduced in Gilbert's report as Figure 49, with the caption, "The East Flank of Mount Ellen, showing the Scrope, Ragged Mountain, Peale, Copper Ridge, and Jukes [Bull Mountain] laccolites, and the Maze and Crescent Arches." The anticlinal Copper Ridge connecting that laccolith with Mt. Ellen stock shows in the drawing. Bull Mountain and Ragged Mountain are bysmaliths.

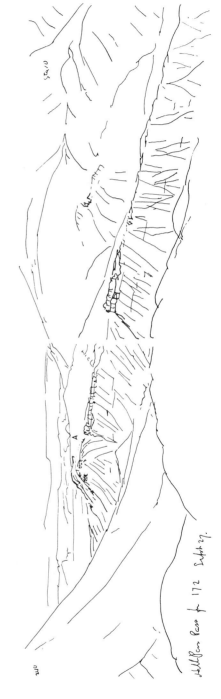

Figure 7.35. "Hill-Penn Pass from 172. September 27." (The sandstone escarpment under "A" is the Emery Sandstone, Gilbert's "C" or Gate Sandstone.)

Figure 7.C. Oblique air photo northwest to Mt. Ellen. Maze Arch in left foreground. Northward, the arched rocks are largely concealed by the pediment gravels deposited by Crescent Creek, which drains from Bromide Basin, site of the Mt. Ellen stock. At the left, over the horizontal ledges of Jurassic formations, is the Copper Ridge laccolith (Gilbert's Sta. 4) exposing its floor and some sills beneath it. The tongue-shaped laccolith forms an anticlinal ridge extending northwestward to Mt. Ellen stock. Gilbert was puzzled by the Maze Arch, and we still are. It has a gentle west flank dipping under the Copper Ridge and other laccoliths of the Mt. Ellen cluster. The structural form is like that of the San Rafael Swell and Circle Cliffs, but very much smaller. No other folds on the Colorado Plateau are like it. It does not seem related to the folding attributed to the intrusions.

Ellen Peak is the high conical peak on the right skyline of Mt. Ellen. Barton Peak, Gilbert's pt. 17 (Fig. 7.4), is the peak on the south side of the mouth of Crescent Creek. It is part of the shatter zone around the Mt. Ellen stock. (Photograph by Fairchild Aerial Surveys.)

{p. 51}   Crossing from 66a to 67a I find beneath the gray slate purple and green earthy hardened shales and then a great mass of quartzite, all dipping 10° to 20° to the north. Then trap.

Point 223 on the flank of Pennell now develops a purple color (pp) in its lower strata (Fig. 7.36) which I do not understand. It is rather a trap than a shale color.

The rock of the station is quartzite throughly metamorphosed and traversed by quartz veins. Perhaps it is Gray Cliff group. [*This is shatter zone on the NW side of the Hillers stock.*]

The rock on 67a [*Fig. 7.29*] is fairly a cross between that of 67 and typical trap and marks the trap as metamorphic even though eruptive. On 67a there is not a trace of bedding but a faint green lamination tho xx are fully developed [*Epidote zone?*].

The crag 189 (Fig. 7.30) shows bedding and the spur leading down from it has red and white sandstone below.

Wind. Yesterday and today the wind has been by day up the cañon (N) and whenever I have been in camp (night and morning) down (S) cañon. The wind on the peaks has been S both days and the clouds have moved from the east (Fig. 7.37).

{p. 53}   **Friday, September 29.**
Camp 11, 7.30 = 22.34 = 8940
Camp 12, 9.30 = 23.60 = 7440

We pass a spring at 8.17, 23.06 = 8070 and come to the level of the lower sandstone bluff 8.25 at 23.24 = 7860.

There is almost no dip here. What there is is a combination of the rise toward Abajo and the rise toward Hillers. A little nearer Hillers the sandstones rise quickly to the Half Dome, getting a dip of 20°.

At 8.40 we cross a marshy creek at 23.49 = 7570.

The camp stream we reach in 13 minutes. At base of 87 = sta. 14 we leave our horses. 24.22 = 6710.

Figure 7.36. Purple strata (p-p) in lower part of pt. 223 (7-33) on the flank of Pennell. "It is rather a trap than a shale color."

From this station the view toward Hill-Penn Pass shows Gate, Lower, *"Gryphaea"* and Belted sandstones in a series of steps. The Sta is on the summit of the Belted.

B.G. in the Maze exhibits a decided arching of strata. The mass lying north of the arch (p. 56).

The crest of the Belted mono outside the eastern arch is bevelled and at one or two points veneered. It shows (Fig. 7.38)

{p. 55}   especially in the long stretch without break.

The eye catches the dip noted on page 35 (Fig. 7.25) on all the spurs of Pennell to 71 [*on Bullfrog Ridge*] (exclusive) and it is steeper than at the point of the first observation.

It now looks (p. 56) as though there might be no fault at the E. base of 40ᵃ but only a sudden drop at the end of the 39 bulge [*Maze Arch, Fig. 7.39*].

The Belted to the southward is half buried by veneering but easily traceable.

Figure 7.37. "September 28. The Lower Lead Cliff, from Sta. 13." (Sta. 13 is the peak of Mt. Hillers. Lower Lead Cliff would be the Tununk Shale.)

Figure 7.38. "September 29. Hill-Penn pass from Sta. 14=87 pt." (The view is NE to the pass, probably from the Pennell Creek benches.)

The black sheets below it are certainly intrusive and materially aid the proof of the bulge structure.

Station 15 at 4.15 = 23.61 = 7310.

There is a trap sheet back of camp 12 that indicates a 15° rise of bedding toward 62 and 63. The spur east of 62 and the one north are tabular or subtabular in their lower part and mutually harmonious (see p. 57).

In the Belted conglomerate we found immense logs (Figs. 7.39 and 7.40).

Figure 7.39. "Sketch of the Maze, from Sta. 14." Sta. 4 = Copper Ridge; 40a = Ragged Mountain.

{p. 58}   Left in cache September 23

15 flour	1 swing
30 jerk	1 pair hobbles
15 pork and bacon	2 kegs
2 apples	1 pair pants
salt	2 boxes
1 milk	1 pack cover
2 chowchow	1 seamless sack
3 yeast powders	1 sack rocks
4 beans	
1 pack saddle	
1 lash rope	

{p. 59}   (Fig. 7.41)

Saturday, September 30. Camp 12 on the 64 fork [*Speck Creek*] of Trachyte Creek. 6.40 = 23.48 = 7510

Not far from camp I come upon sandstone which proves to be Belted. It rises toward the 62 spur of Hillers and I think it rises

over the trap at *A A A* (p. 57 [= *Fig. 7.40*]) and appears in *B B B* (Fig. 7.41).

I forgot to note yesterday that the curious metamorphic quartzite which appears a little below the *"Gryphaea"* horizon across Trachyte Creek appears also on this side. Is it due to solfatarro action?

All this is confirmed by later view and the sketch on next page shows that the reverse dip is toward Jerry and Jerry Pass. The Jerry bulge and the 59 Bulge may be independent.

Figure 7.40. "September 29, from Sta. 15." (View is SW to Mt. Hillers.)

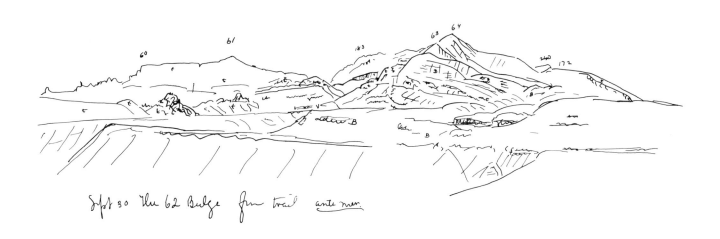

Figure 7.41. "The 62 Bulge from trail. 'ante men' " (See Fig. 7.6). 172 is the Stewart Ridge laccolith. 240 is Bulldog Peak. 60 and 61 are Gilbert's Jerry = Sawtooth Ridge.

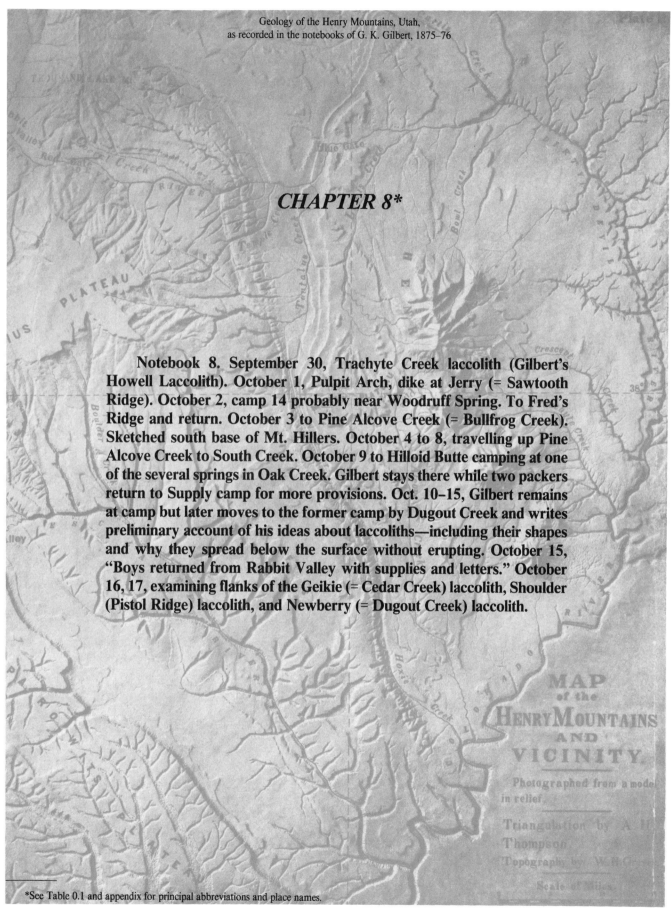

*CHAPTER 8**

Notebook 8. September 30, Trachyte Creek laccolith (Gilbert's Howell Laccolith). October 1, Pulpit Arch, dike at Jerry (= Sawtooth Ridge). October 2, camp 14 probably near Woodruff Spring. To Fred's Ridge and return. October 3 to Pine Alcove Creek (= Bullfrog Creek). Sketched south base of Mt. Hillers. October 4 to 8, travelling up Pine Alcove Creek to South Creek. October 9 to Hilloid Butte camping at one of the several springs in Oak Creek. Gilbert stays there while two packers return to Supply camp for more provisions. Oct. 10–15, Gilbert remains at camp but later moves to the former camp by Dugout Creek and writes preliminary account of his ideas about laccoliths—including their shapes and why they spread below the surface without erupting. October 15, "Boys returned from Rabbit Valley with supplies and letters." October 16, 17, examining flanks of the Geikie (= Cedar Creek) laccolith, Shoulder (Pistol Ridge) laccolith, and Newberry (= Dugout Creek) laccolith.

*See Table 0.1 and appendix for principal abbreviations and place names.

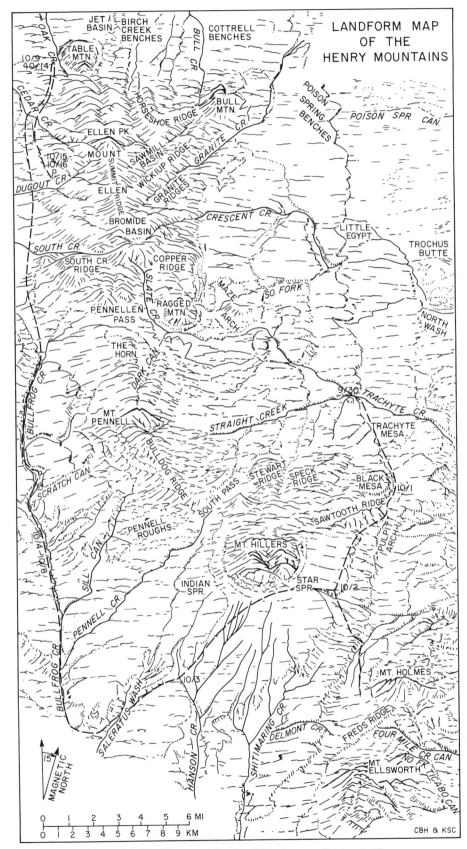

Figure 8.A. Landform map showing Gilbert's route October 1–17.

Figure 8.B. Oblique air photo looking south along Mt. Ellen to Mts. Pennell (right) and Hillers (center). The trap-door structure that characterizes the Henry Mountains bysmaliths is well displayed at Table Mountain in the lower right. This is Gilbert's Hilloid Butte—Marvine laccolith in his report. The sandstone roof over the intrusion is Morrison Formation faulted upward above the Tununk Shale and Ferron Sandstone, which are turned up steeply and cut off discordantly against the semi-circular trap-door fault along the near side of the intrusion. This marks the steep, bulging, distal end of the intrusion. The steep dips flatten southward to the hinged proximal side which consists of an unbroken arch of the Tununk Shale and Ferron Sandstone and forms the anticline extending southward under the intrusions forming the high north end of the mountain. It would appear that the feeder laccolith broke upward to form the bysmalith where the feeder extended northward beyond the limits of the overlying, higher laccoliths. (Photograph by Fairchild Aerial Surveys.)

Figure 8.C. Pack train of the 1936 U.S. Geological Survey field party crossing the Trachyte Mesa (Gilbert's Howell) laccolith. This view, about east, is almost the same as the one drawn by Gilbert, Figure 8.2. Entrada Sandstone is arched upward at the north edge of the intrusion.

G. K. GILBERT, POWELL SURVEY, P.O. BOX 806, WASHINGTON, D.C.
OPENED AT CAMP 13 ON TRACHYTE CREEK NEAR STA. 13

{p. 1}   **Saturday, September 30, 1876** contd from notebook 8 [*should be 7*]

At the west end of 149 [*249?*] the relation of the trap and the Vag. is visible. The Vag. arches over unbroken.

The sketch on the next page [*2*] does not well express the facts. [*The sketch on p. 2 is omitted because the view is almost duplicated by the much better one on p. 3, Fig. 8.2.*] A section is better (Fig. 8.1). The sketch on p. 3 (Fig. 8.2) is a little better. The sandstone and shale are little altered. The shale is somewhat hardened and the color is discharged changing red to white along the lower line of contact of the upper sheet. The trap is very dark.

I cannot tell whether these sheets are connected with Hillers. They dip with the bedding to the west and disappear. The sketch on page 3 represents about 40 ft height. Camp 13 at base of pt. 250. 30 ft above Trachyte Creek.
1110 = 24.85 = 6010, 250 = sta. 16 = 11.30 = 24.54 = 6480.

{p. 4}   Sta 16, pt. 250 is a veneered butte near the bend of Trachyte Creek. Under the veneering is Vagabond exhibiting considerable very irregular flexuosity. It covers a broad area to the N projecting at several points through the veneering. The most notable is 246. Southward it is largely concealed by the veneer.

Figure 8.1. Section showing arching of Vagabond Sandstone (= Entrada) across the intrusive sheet at Trachyte Mesa.

The trap sheet 249 is only 30 or 50 ft thick. We see its eroded SE face, resting on Vag. sand. It shows more prismatic structure throughout than I have seen elsewhere in the trap.

59 [*Black Mesa*] is a massive sheet about 500 ft thick dipping N (by NE) at a low angle (10°) or away from the Jerry [*Sawtooth Ridge*] dike (60). I see strata rising to it at one point only and that at the right tangent.

Just below camp 13 the creek cañons into what I take for G.M. If it is, then the foss. Jura is entirely wanting. Indeed I have never seen it about the Henry Mts. At any rate the rock mentioned constitutes pt 198, 199, 200, and 207.

To carry out the following plan of triangulation, it is desirable to follow this order.
1st, visit 160 and, ascertaining that is is visible from 214 and 1, define it by a monument. [*159 in Fig. 8.3.*]

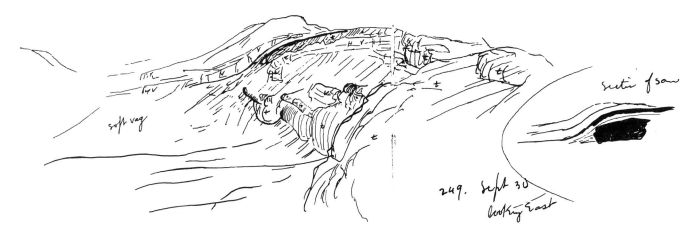

Figure 8.2 Gilbert's sketch and cross section of the north flank of the intrusive sheet at Trachyte Mesa. View east of Gilbert's pt. 249 (or is this pt. 249?). See also Fig. 8-C.

2nd, climb 220 and clear of trees.

3rd, 214.

4th, 246.

5th and 6th, either 73×1 or 1×73.

I think no angle of this system will be less than 30°.

Sunday, October 1. Camp 13, in Trachyte Creek. 7.5 a.m. - 24.89 5970 (30 ft above creek).

Sta. 17 is SE of 59 - 9 a.m. = 24.40 = 6520.

The 60 dike [*Sawtooth Ridge*] is a broad one and is encased in Belted [*Morrison Formation*] conglom. which is nearly level and rather synclinal than otherwise.

{p. 6} It is entirely separated from 59 [*Black Mesa*] by sedimentaries. The sandstone and trap interlock along the side of the dike (Fig. 8.4)

248 [*sill east of Black Mesa*] is precisely similar to 249 [*Trachyte Mesa*] with the addition of dikes. One of the latter is beautifully cut across by a superimposed drainage line. The prismatic joints have conspired to make a cleft narrow as Pfeffer Gorge [*in New York?*]. There is a block of Vag. on top of 248.

The Pulpit Bulge is no more sketchable than the Maze, which it resembles. Its rock is the G.M. with the Vagabond around the edges, and all are brick-red. Beyond it the distance is short to the G.M. which is level or rising eastward.

The Pulpit rock is G.M. and level. The dome extends to the base of 59, to the end of the 60 dike, almost to 248 and 247 and eastward a half mile. Southward I do not see its limit.

{p. 7} I cannot make a location on sta. 17. The trouble about Jerry is involved and was to some extent on sta. 16.

Noon camp at the head of a veneered slope up which we have climbed 2 or 3 hours on a trail. [*Evidently the Indian trail from Trachyte Creek to Star, Woodruff, and other springs SE of Mt. Hillers.*] At 11.5 am read 23.58 = 7450. In climbing we have

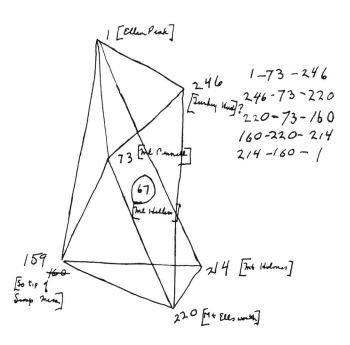

Figure 8.3. The triangulation net discussed by Gilbert on p. 4, 5. Along the edge of the page showing these stations he wrote, "From each of these stations four of the others are visible. The places out of sight of each other are 1-220; 73-214; 159-246."

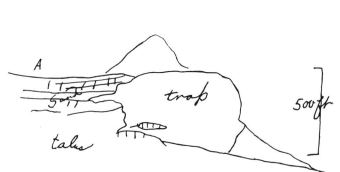

Figure 8.4. Cross section of the east end of Gilbert's "Jerry" (= Sawtooth Ridge).

Figure 8.5. View west along the crags formed by the sandstones turned up against the south side of the Mt. Hillers stock. View is from hill above Starr Ranch campground, location of his camp 14 (p. 9).

passed a sheet of unknown extent in the Vag. ½ mile south is an outcrop of Belted dipping 20° to the south. We are not much above the level of Belted supporting the 60 dike.

We leave horses at 12.15 on a spur of Jerry which is itself a branch dike. 23.24 = 7850.

So far as we have climbed (8000) we have shale in sight in place and I think shale immediately underlies our noon camp. Here we have its contact with the trap and find it hardened. At a little distance it is soft.

Jerry at 1.5 = 22.27 = 9,000
 3.50 = 22.22

There are dikes radiating from Jerry. One is No. 60, another is toward Jerry Pass. A third bears toward the SE Spur of Hillers. The fourth we climbed and can trace straight and thinning to the S flank of the Pulpit Bulge [*E base of Mt. Hillers near Highway 276*].

The 62 bulge [*North Sawtooth Ridge laccolith*] ends NE-ward just at the top of 59 [*Black Mesa*] and to ' ' ' '
' '.

{p. 8}   In the Mt. spur there is a dike down the middle and I can see level Belted high up resting on 1000 ft of trap. This way is dip at a high angle (say 25°) of the shoulder of the spur under 62 and 63, is chiefly floored by Belted con[*glomerate*]. It forms the pass rock and follows the creek down to the left. The 187 [*189?*] spur shows though nearly concealed by trap debris from dikes and in spur 184 it is well tilted up. The 62 bulge [*North Sawtooth Ridge laccolith*] merges with the great Hillers bulge [*at the shatter zone*] and is only a spur of it.

The Pulpit Bulge looks quite symmetrical but the Pulpit is not in its center by ¼ ' '. It is to the north, and south of the center is the big red castellated butte.

The summit of Jerry is half trap and half quartzite.

Something is wrong about my plane table work. [*I know the feeling.*] Either I have made some error in location that has accumulated or there is some error in the positions on which I have based. Now I am going to determine a new position for Jerry based on the line from Hillers to Ellsworth.

Barometer gone to smash. Ditto shovel.

Figure 8.6. October 8. Ellsworth as seen from Gilbert's pt. 217 (Sta. 19, probably on escarpment of Morrison Formation about west of Mt. Holmes).

Figure 8.7. View east to Mt. Holmes (= Gilbert's H.V.) from Sta. 19. The ridge on the skyline at the right is the Theater laccolith that was injected southward from the Mt. Holmes stock.

{p. 9}   Camp 14 is near one of our last year's camps. The one to which we returned from Ellsworth.

Monday, October 2. Camp 14 to pt. 217 [*Fred's Ridge below Ellsworth and Holmes*] and return. Saddle below Belted, 8.30 = 24.33 = 6600.

Saddle between 217 and 218, 9 am = 23.61 = 7410.

Sta 19 [*no sta 18 in notes*] = pt 217 = 10 am = 23.20 = 7900. At 3.20 = 23.18.

{p. 13}   **Tuesday, October 3.** Camp 14. 7.5 am = 23.50 = 7560.

The general structure of Ellen and Pennell is a system of bulged intrusions of trap compiled as irregularly as the secondary cones of a volcano. Successive jets of trap finding passage at diverse points ceased their upward movement at points equally diverse and spread in lenticular forms lifting the superior strata.

Camp 15, on Pine Alcove Creek [= *Bullfrog Creek*] at the lower sandstone [*Ferron*] 1.30 = 24.73 = 6140.

There is a thin bed of coal on the sandstone, covered by the shale, where Dinah Creek cuts it, it is thicker, perhaps 3 feet. [*This is at or near the Stanton Coal Mine, developed in the 1890s to supply coal to gold operations on the Colorado River.*]

Sta. 19 bet. Ells and H.V. [*Mt. Holmes*] is on a dike not connected with either mountain.

Wednesday, October 4. Camp 15, 6 am = 24.55 = 63.50

At 8.50 we leave horses at base of 159. 24.31 = 6610.

Gray shale to 23.90 = 7080, then the same alternating with sandstone to 23.86 = 7130 = 9.13 am

Massive sandstone to   . [*Left blank.*]

Above this are 100 feet of shaley

{p. 14}   sandstone and another equal bed of massive sand-

Figure 8.8. Sketch of the south base of Mt. Hillers showing the crags of the sandstones turned up at the edge of the main intrusion, now known to be the stock at the center of the Hillers cluster of laccoliths (cf. Fig. 3B).

stone which we do not climb but go around rising with the dip to

9.50 = 23. 42 = 7640 = sta. 20 = pt 159 at 12 m = 23.38
$$12.40 = 23.38$$
$$3 \text{ pm} = 23.34$$

Our journey is the old trail along the top of the lower sandstone. The sandstone is not thick and midway it so nearly fails that the cliff is easily ridden over at many points. There is a slight flexure of the strata which carries the mono N at the same point. It is a faint mono fold with NE throw and merges in the Gate Cliff with the Waterpocket fold. The two give 159 [*S tip Swap Mesa*] a dip to the NE for ¾ mile and 158 shows in it.

The Waterpocket fold surely brings up the Carb. and lays bare several hundred feet of it.

Thursday, October 5. Camp 15.   7.30 am = 24.62 = 6260.

We follow up Pine Alcove Creek. Dinner camp is a little above the Pass of the Gate [*Emery*] Sandstone, 11.30 = 23.83 = 7160.

Top of the sand rock = 23.73 = 7280.

Camp 16 (3 pm = 23.35 = 7720; 3.15 pm = 23.29 = 7790) is at point 114. Two miles below, the creek (P.A.) runs in a narrow box cañon in the Gate Sandstone. We found it convenient to travel outside the cañon.

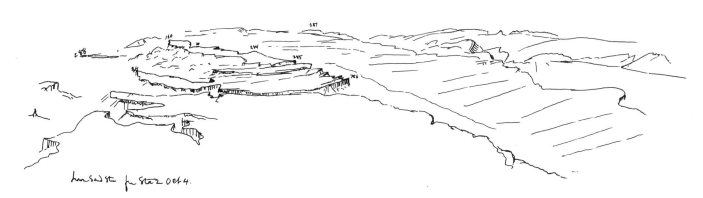

Figure 8.9. "Lower sandstone from Sta. 20." There is uncertainty about the location of Station 20 [= pt. 159]. Using bearings in Figure 8.10 as well as in this one, the station appears to be on the butte of Emery Sandstone next south of the one identified as Cave Point on the Mt. Pennell quadrangle. Gilbert's "Lower Sandstone" would be the synclinal point of Ferron Sandstone along the east side of Bullfrog (= Pine Alcove) Creek.

Figure 8.10. "October 4, Sta. 20." See also Figure 8.9. This is view NE to Mt. Pennell. The hogbacks dipping off that mountain are formed by intrusive sheets in the Bluegate Shale just above the Ferron Sandstone. The bulge at the north base of Mt. Pennell is the Horn Laccolith. Left of that is the profile of Mt. Ellen showing Ellen Peak and the north and south Summit Ridges. Right of Pennell is Mt. Hillers with Bulldog Peak at its north base and the crags of sandstone turned up against the south base of that mountain (see Fig. 8.8).

{p. 17}   The creek barely touches the Pennell rise of the rocks but below here runs chiefly in the Plateau proper. The Gate Sandstone barely sinks below the creek below 114. The full stream on which we lunched comes from the Pennell Branch.

Friday, October 6, 1876. Camp 16 = 6.45 = 23.32 = 7760. Sta 21 = pt 74. 9.30 am = 21.00 = 10590.

The monoclinal at the W base of the Ellen and Pennell Swell is flexuous having westward convexities at 118 and near 115 [*Emery Sandstone*] and eastward at 116 and again this side of 113—along our trail of this morning. The latter is perhaps to be regarded as the groin bet. Ellen and Pennell. Still structurally E [*Ellen*] and P [*Pennell*] are not the distinctive units that the other three are [*see Fig. 0.D*]

The rise toward Ellen (toward spur 26) is continuous so far as I can trace it; but toward Pennellen Spring it diminishes at one point to nothing and then increases again. Toward Pennell the rise is one of increasing steepness to this point. In ascending we followed the dip of the Gate Stone more than half

{p. 18}   way and then descended nearly to the lower Blue and to a trap dike that forms the point.

Saturday, October 7. Camp 16 7 am

Ride to base of sta. 22, 8.30 = 23.41 = 7650

We have been riding for some distance over arenaceous shale. The same continuous with occasional interruptions by sandstone to (8.40) 23.12 = 7990. Then there is little but sandstone to the top.

There is a coal horizon not far above the Gate Sandstone but nothing economic.

Sta. 22 = Pt. 105 = 8.50 = 23.04 = 8090, is not on the summit of the sandstone, which rises about 100 ft higher. Sta. 22 = 8.50 = 23.04 = 8090

        9   = 23.03
    11.10 = 23.01
    12.45 = 23.00

The main body of this upper plateau is level. It is slightly involved at the edges in the Waterpocket and Henry Mt. foldings. The south Gate Plateau seems to have a slight general rise southward. The end of 159 [*S tip of Swap Mesa*] joins slightly in the upward turn of Point Retroussie [*Flap of hat*] but the salient nearer Pine Alcove Cr. has a downward curve at its south end. Everything east of the creek rises to Pennell and between 143 and 161 there is a bend [*east side Cove Flat, between Scratch and Pipe Spr. Creeks*] which horseshoes

{p. 20}   into the Pennell slope. The western edge, as already known, has a slight upturn throughout.

Travel is difficult if not impossible under the south edge of the upper Plateau.

There are some points of interest about the drainage of the NW cañon of Pennell, the one which passes just north of Sta. 21 [*Pine Spr. Canyon*]. Down to where it leaves the trap it is deep but just there it is shallow and it has had thence a broad delta-cone now marked by patches of veneering which are fractions of

concentric cones and which appear on 300 and almost down to 113; on 299 and the slope below it, on 297, about the base—even the south base of Sta. 21; and all the way down Pine Alcove Creek to 161 (beyond?).

The prolongation of the Gate sand in 300 is due to the protection of the delta cone.

It may even be true that the drainage of the cañon [*see Introduction under Land Sculpture*] has been once carried [*NW*] to Cache Creek. Herein in manner of shifting drainage which ranks with the retreat of cliffs and with ponding—shifting on an inland delta cone.

100 ft filling of the present channel would through [*throw?*] the water the other side of 299 and still further filling would send it beyond 300 and around 113 via 114. Further action might carry it to 117 and so to the Dirty Devil River instead of the Colorado.

In the destruction of successive cones, water courses have been stranded at several points. The 114 channel [*canyon at Stevens Narrows*] is one instance, 299–300 is another, and one which heads at 298 is a third (see p. 21). See also Fig. 8-D.

{p. 22}   Sta 23 is between 105 and 104. Little to be seen new. At 3 pm 22.85 = 8300.

This is practically the top of the sandstone. Think there is a slight dip from Pennell or from Sta. 22. There also is a dip from 22 and 23 to the N. Perhaps the Plat. is syn. both ways.

Note. The Mts are rugged in the order of their smallness. 1st HV, then Ells., Hill., Pen., Ellen. It is probably a matter purely of climate.

Descending I hunt for fossils and find not even a fucoid—nothing but coal and little of that. The shale is on the whole more arenaceous than the Blue Gate. Horses again at 3.50 = 23.34 = 7700. Camp 16 at 5.10 pm = 23.34 = 7850.

Sunday, October 8, 1876. Camp 16, 7.10 am, 23.30 = 7790.

Divide between Pine Alcove and Very South Cache Creek 8.40 = 22.99 = 8150.

V[*ery*] S[*outh*] Cache Creek near 118 = 23.17 = 7030.

The divide is close to V.S.C. [*Very South Cache Creek*] and is slowly working farther.

{p. 23}   Saddle above 22.21 = 9080

We come upon S Cache Cr. at 22.65 = 8560.

Cache Camp 12 M, 22.90 = 8250.

Monday, October 9, 1876. Cache Camp, 9.45 = 22.95 = 8200. Camp 17 at Base of Hilloid Butte [*Table Mtn.*].

1.15 = 22.48 = 8750

6.10 = 22.47

I had begun to suspect that the Hilloid Butte was not so put up but was a mere lump like NE [*Bull Mtn.*] but approaching it I find not mere trap sheets but sandstone strata leaning against its west side. The dip low down is at a high angle but I think the angle diminishes upward.

The lumpy character of the whole flank we have passed is

168

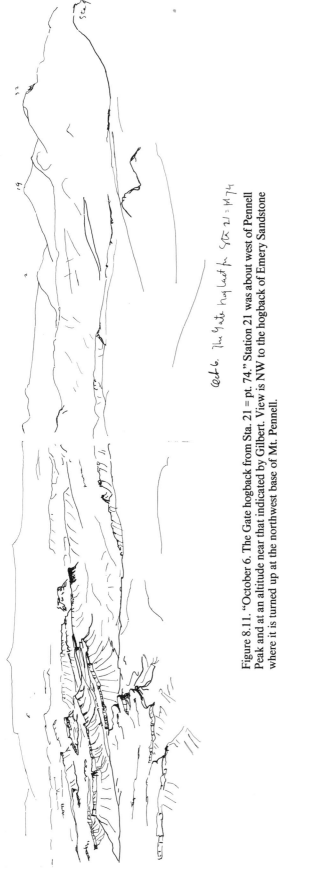

Oct 6. The Gate highlandt for Sta 21 = pt 74

Figure 8.11. "October 6. The Gate hogback from Sta. 21 = pt. 74." Station 21 was about west of Pennell Peak and at an altitude near that indicated by Gilbert. View is NW to the hogback of Emery Sandstone where it is turned up at the northwest base of Mt. Pennell.

Oct 7. Pennell from Sta 22.

Figure 8.12. "October 7. Pennell from Sta. 22." Station 22 is on the Mesaverde Sandstone at a point along the east edge of Tarantula Mesa. The view is a little south of east to the back side of the hogback formed by the Emery Sandstone (Gilbert's Gate sandstone) at the NW base of Mt. Pennell. The hogbacks inside the Emery Sandstone are formed by sills (cf. Fig. 8.10).

Figure 8.D. Oblique air photo looking up Cedar Creek draining west from Ellen Peak, the conical knob on the skyline right of center. The creek still is perched on the gravel veneer covering the pediment eroded in Mancos Shale. At any time the creek is going to become diverted northward off the gravel-covered pediment to the lower gullies eroded in the bare shale north (left) of the pediment. When that happens the creek will deposit its load of gravel on the bare shale pediments north of the high gravel veneered one, which—by sapping around the sides—will become one of Gilbert's isolated "hills of planation" (see Fig. 63 of his published report). (Photograph by Fairchild Aerial Surveys.)

evident but the details can better be described after visiting other points. The chief rock exposed at heads of taluses is Belted.

This morning Elisha and Frank started for Rabbit Valley for supplies with

{p. 24} three pack mules.

Climbing a hill near camp I see the west flank of the Hilloid and make the section in Figure 8.13. I interpret *a* and *B* by the color of the intervening shale.

Spur 2 I make little of. Its sides are great slides of trap. At one point on its NW base is a slab of trap in place *a* in Figure 8.14, that seems to indicate a steep dip and just outside it is a sheet *b* that inclines only 10° from the spur.

The water which rises at our camp runs only a short distance. It probably rises again below. [*It does rise again, at Willow Spring a mile north.*]

Tuesday, October 10. Camp 17. 6.50 am = 22.45 = 8800 Showers yesterday pm and evening; at 22.27 = 9000 across the *"Gryphaea"* beds is a

{p. 25} little sandstone coating a heavier coarse conglomerate. The dip is 45° and the strike is toward the red gate or

Figure 8.13. Section on west side of Hilloid Butte (= Table Mountain) as seen by Gilbert from a hill near camp 17. "*a "Gryphaea"*; B, Belted [= *Morrison Fm.*]; and *T T* trap."

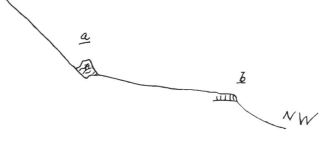

Figure 8.14. Section on spur 2 described on p. 24.

tangential to the butte at this point. This looks as though the dip was circling in toward the Hilloid-Ellen pass.

At 7.35 = 22.03 = 9220. I am on a hackly gray slate having passed more of same but not much as the grade and the dip are about equal, and the strike is working around into the mountain. The dip here seems 45° and the strike toward Tantalus point. Looking northward along the slope I get a profile of Belted Conglomerate dipping beyond the stable angle (Fig. 8.15), say about 60° and striking directly from me and toward Mt. Alice, or the middle of the South Gate post. Just above me is trap in place.

A little later I find a large block of slate, presumably in place, dipping 60° toward Ellen (Fig. 8.16).

We reach the N end of the

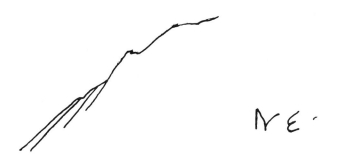

Figure 8.15. "Belted conglomerate [*Morrison Fm*] dipping beyond the stable angle on the west side of Hilloid Butte [= *Table Mountain*]."

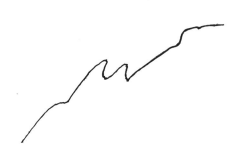

Figure 8.16. "The trap on the profile at the N has a dipping bedded look on the outside and then a massive look within."

{p. 26}   crest at 21.70 = 9740 and the summit (sta. 24) at 8.25 = 21.48 = 10,000 [*USGS Mt. Ellen quadrangle gives summit as 8521*] 2.33 = 21.39

I am surprised to find slate with "*Inoceramus.*" It lies horizontal. Yesterday I noticed that the cap north from the crest seemed bedded with a slight dip north. The eastern slope is not so steep as the western, and the shoulder which is 3–400 feet lower

exhibits a sheet of trap level on top and curving down to the east and to the north. Beyond it is a valley in the Lower Blue Shale [*Jet Basin*]. Beyond that again (NE) are buttes of uprising lower sand [*Ferron Sandstone*]. I fancy the same sandrock underlies the trap sheet. In that case this slate on top is Gate.

Going to the west edge I get a view of the sandstone [*uppermost Morrison Sandstone, see Fig. 8.B*] capping the north end of the butte and curving over regularly to the N and W. It is probably the lower sandrock and it rests (in part at least) on trap. The same sandstone forms a little butte at the NW base beyond the creek with a dip of 20° or 25°, half from the butte and half from the main mountain.

Camp 17. 3.45 pm   22.38 = 8900 (Figs. 8.17 and 8.18).

{p. 29}   **Wednesday, October 11.** Camp 17. 8 am   22.44 = 8820.

A shower last night. Snow on M.L., Aquarius, and Hilgard. None on Ellen.

Riding to Point 328 (Fig. 8.18) I find it is good for a station. There is a dip of 6° or 8° from the Mt. of Gate Shale and a little way back (½ mile SSE) we crossed the lower sand barely revealed through the veneering. I read at 9.45.   23.66 = 73.60. Descending eastward into a wash in the badlands I read 23.92 = 7050.

Station 25 (= 10.50 am = 23.15; 12.45 = 23.10 = 7960 is on a veneered slope not very far from camp 17. It is chosen to command a view of the Hilloid Butte at its N end. The smooth curves of the top are all structural although the sandstone cannot be traced very far. The flying buttresses are also structural, I am confident—although I see no sedimentaries beyond those noted yesterday and although there is an obscurely horizontal streaking of the trap. The buttresses are too slabby to be due to anything but interbedding or a structure imposed by contact.

It now looks as though 318 might have slidden. There are other fragments of that sandstone lying at various angles.

{p. 30}   I suspect the vertical in this drawing is too small for the horizontal. It can be tested by the angles from sta. 25.

{p. 31}   The trap on the veneering is corroded (by the shale?).

Crossed a creek that flows from the big dry gulch we crossed yesterday—at 22.98 = 8170.

Point 313 = 22.45 = 8800 at 3.15

Camp 18 is at the base of 313 (Fig. 8.17) on a small creek flowing down the veneering. 5 pm 22.66 = 8600.

Rain.

Thursday, October 12. Camp 18 at the base of 313 [Fig. 8.17].

7.45 am	= 22.60 = 8610
9.15	= 22.65
9.55	= 22.63
11.10	= 22.63
12 m.	= 22.65
3 pm	= 22.61
5 pm	= 22.60
6.10	= 22.56

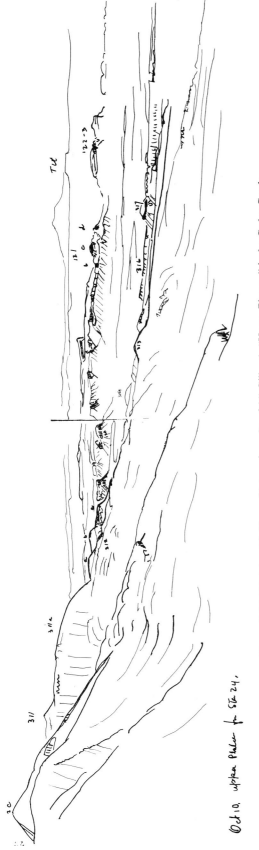

Figure 8.17. "October 10. Upper Plateau from Sta. 24." Gilbert's "Upper Plateau" is the Cedar Creek bench; his Sta. 24 is top of Table Mountain.

Figure 8.18. "October 10. N. Gate Plateau [*Stevens Mesa*] from Sta. 24 a." Station 24 is top of Table Mountain, Gilbert's Hilloid Butte. Stevens Mesa, formed of Emery Sandstone, lies along the axis of the Henry Mountains syncline. It is divided by Sweetwater Creek which flows right (north) under Gilbert's points 326, 325, 323, 322 and emerges from the mesa at d.

Figure 8.19. "Hilloid Butte [*Table Mountain*] from Sta. 25." Station 25, about a mile northwest of Table Mountain, is on Ferron Sandstone. The crags around the base of Table Mountain are a chilled facies of the diorite porphyry. Between the instrument station and Table Mountain is the valley of Oak Creek, with a spring (Willow Spring) used by Gilbert. See also Figure 8.B.

Figure 8.20. Unlabeled drawing on p. 31 of the notebook.

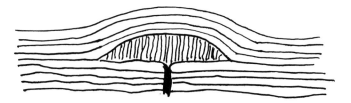

Figure 8.22. "The simplest type of Henry Mt. structure is a lenticular mass of trap above which the strata are arched." Notebook 8, p. 34. This records Gilbert's discovery that intrusive rocks deform the host rocks into which they are intruded. Later structural mapping has shown that there are no feeder pipes beneath the bulges; rather, each is the distal end of an anticlinal bulge, the axis of which, in all cases, extends back toward the central stocks. Compare Figure 8.B.

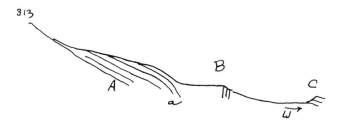

Figure 8.21. "The Belted series [*Morrison Formation*] juts in *313* above the veneer and the same rock appears in the bluff at A. At *a* they curve downward and at B is *"Gryphaea"* series with a reversed dip at top. Farther west at *C* the lower sandstone [*Ferron Sandstone*] rises toward the Mt. with gentle dip, and I could see after climbing the cliff that between B and C was another mass of Lower Sandstone considerably disturbed but dipping in general either 30° to the east or 150° to the west. It remains to determine whether the phenomena are of the landslip order and superficial or deep-seated and bared by erosion. I saw no trap."

Work was stopped yesterday by rain which began about 4 pm. There has been some intermission since, but this morning we are at the lower edge of a cloud and in a drizzle of rain half the time. Two grand showers yesterday moved from Aquarius Plateau across the Waterpocket fold and the Blue Gate.

No plane table work was done on sta. 25 yesterday because the mts. were covered by clouds and a location was

{p. 32}   impossible. It can be placed on the map after points have been fixed at the east 121d, 319, 327 etc.

Spherical excess. The excess of a spherical triangle is proportioned to its area. The excess of a right angled spherical triangle is 90°. The area of a L spherical triangle equals ½ a great circle's area. The area of a great terrestrial circle (approx.) is π (4000 miles)^2.

$$\therefore \frac{\pi}{2}(4000)^2 : A :: 90 \times 3600 : E$$

in which A = area of a terrestrial triangle in square miles, and E = the spherical excess of the same in seconds of arc.

$$E = \frac{90 \times 3600 \times 2}{\pi (4000)^2} A = 0.0129A, \text{ or } A \ 78E$$

That is, a triangle with an area of 78 square miles has a spherical excess of one second.

The largest triangle in the Henry Mts. measures about 23.5 × 19.2 × 19. Its area is about 177 sq. miles and its spherical excess

$$E = \frac{177}{78} = 2.3 \text{ seconds.}$$

The excess of the whole polygon of triangulation is not greater than 5″ and is quite inconsiderable as compared with the probable error of the measurement of our angle—1 minute.

In approaching 313 yesterday a curious structure was noted in the N-facing cliff, but the approaching storm prevented a careful examination (Fig. 8.21).

The Gate Sandstone all the way from 317 to 318 rises toward Ellen, 317 more than 318. The foot of the fold is not far beyond those points, but there is a very gradual descent to about Cache Creek cañon—say 325, 322 [Fig. 8-18] and 320d—where the Gate fold begins.

{p. 34}   *In general.*

The simplest type of Henry Mt. structure is a lenticular mass of trap above which the strata were arched, the section being as in Figure 8.22. In the Hilloid Butte the curved strata are to be seen on top and on two of four sides. The third side is masked and from the fourth they have been eroded; the base of the trap is not seen.

In the NE Butte [*Bull Mountain*] none of the curved strata are visible, the whole summit is trap. But the level beds below are visible on nearly every side [*see Fig. 6D*].

In the Bowl Creek Bulge [*Bull Creek laccolith*] the curved strata above are retained except at one point where they are cut through and the upper part of the trap mass is exposed.

In the Station Three Bulge [*Wickiup Ridge*] much is masked, but portions of the trap mass and of the upper and lower strata seem to be exhibited.

In the Crescent, in the Maze, in the NE Arch, in Pulpit Arch, the upper curving strata alone are exposed.

In the 59 [*Black Mesa*] are seen the lower level beds, the trap mass, and at a single point the uprising strata at the edge of the trap. The top of the trap retains, as in the Hilloid Butte, its original shape and may have strata.

In 249 [*Trachyte Mesa*] every element of the type, except the supply dike or chimney, is clearly visible. [*The feeder to the Trachyte Mesa laccolith is the famous (or infamous) cactolith. See introduction section on forms of intrusions.*] At one end the upper strata complete the

{p. 35} arch; at the other they have been completely removed and the trap can be seen resting on level strata.

In 248 [*sill east of Black Mesa*] the exposure is the same as in 249.

In the Station 4 [*Copper Ridge*] head a section is shown across the trap mass and the underlying undisturbed strata.

In the 40 Butte [*Ragged Mtn.*] the trap is shown resting on the lower strata which here are bent by other and lower bulges [*but see Fig. 7.21*].

In Sentinel Butte [*The Horn*] the lower rocks are obscurely indicated and the form of the trap mass.

The *form* of the trap mass is never fully shown but it can be described in a general way in several cases. 249 and 248 are thin broad sheets with one diameter greater than the other. 59 [*Black Mesa*] is thicker and the Hilloid thicker still but both rather flattish. In NE Butte [*Bull Mtn.*] and in 40 [*Ragged Mtn.*] the vertical dimension must have been almost equal to the lesser horizontal. The Hilloid [*Table Mtn.*] is steep on one side and of gentle dip on the other resembling in this Mount Hillers. None seem so nearly symmetrical as Mt. Ellsworth.

Accessions of the trap mass are found of 3 kinds.

1st, there are sheets parting the curved strata above the main mass and conforming with them.

2d, there are sheets parting the inferior strata, and conforming.

These two phenomena may be regarded as attenuated forms of the typical trap mass.

3d, there are dikes traversing the strata in various directions and not conforming. They usually radiate from centers of bulge

{p. 36} injection.

Examples of arching sheets are found associated with the Hilloid Butte, the Bowl Creek Arch, the Sta. 3 Arch [*Wickiup Ridge*], and the Cache Creek [*Dugout Creek*] Arch, the great Ellen Arch, the 247 sheet [*east of Black Mesa*] the great Pennell Arch, Mt. Hillers. Mt. Ellsworth. and H.V.

Dikes radiate from Jerry, from the NE Butte, from the core of H.V., from the core of Hillers, from the core of Ellsworth, from 248 [*sill E of Black Mesa*], and from the Great Ellen and Great Pennell swells. They show above the main trap mass in the Crescent, in Jerry, and in H.V. and in Ellsworth.

Injections are combined and *grouped* variously. A trap mass with sheets above and below may be regarded as a system of exudations from one chimney. Sometimes the masses rest on the same floor; as for example, the Hilloid, the NE Butte [*Bull Mtn.*], the Sta. 4 [*Copper Ridge*], the Sta. 3 [*Wickiup Ridge*], the 59 [*Black Mesa*], and the 40 [*Ragged Mtn.*]—all on the Belted Conglomerate. Sometimes they rest on different floors. 248 [*Trachyte Mesa*] and 249 rest on Lower Vagabond. The Bowl Creek Bulge [*Bull Creek laccolith*] lifts the floor of the NE Butte. The NE arch is cut to the GM without reaching trap. So are the Crescent, Maze, and Pulpit arches, and H.V. [*Mt. Holmes*], while the floor of Ellsworth is lower than the upper Aubrey. On the other hand Sentinel Butte [*The Horn*] probably has a floor of lower Blue Cliff. Bulges at different levels overlap. The Maze arch is compound in itself and it is overlapped by 40 [*Ragged Mtn.*] and Sta. 4 [*Copper Ridge*]. The

{p. 37} [*word missing*] adjoins the Maze and the NE arch and is overlapped by the Sta. 3 [*Wickiup Ridge*] bulge, the Bowl Creek arch, and the NE Butte. The latter overlaps not only the Crescent but the NE arch and the Bowl Creek arch. There is a fine combination yet to be sketched of which the Cache Creek is a member. The Pulpit, the Jerry, the 59 [*Black Mesa*], and the Hillers arches adjoin at different levels and the 248 [*Trachyte Mesa*] may run under the 59. Ellsworth and H.V. adjoin *aux grainese*. Ellen and Pennell each seem to be a congeries of minor trap masses rather than an individual.

In fine, the lavas rising in this district have not invariably followed the same flue but have divaricated in their upward courses just as have the lavas of other volcanic districts recent and extinct. But instead of rising to the surface and piling up mounds at their several vents they have stopped beneath the surface and piled up subterranean mounds over which superficial hills must have been produced by arching. Moreover they have made these subterranean mounds at many different levels so as to produce a structure of which Figure 8.23 is an ideal cross section.

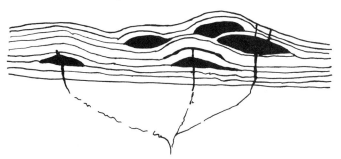

Figure 8.23. Gilbert's idealized cross section of the clustered intrusions and arches over them in the Henry Mountains. Compare with Figure 10 in his monograph.

{p. 38} Contemporaneous and subsequent denudation have left the existing phenomena.

Why have Henry Mt. lavas spread beneath the surface when many other lavas have risen above it? What is the common law of the two phenomena? I conceive that it is the simple hydrostatic one, that a fluid tends to rise above other substances which are denser than it, and to sink below such as are less dense. Molten rocks which are of less specific gravity than the walls which contain them tend to rise and (having vent) do rise until a hydrostatic equilibrium is produced. This will occur, if the covering rocks cohere, when the product of the mean density into the depth (= the weight for a unit of area) is equal for fluid and for solid cover. If the covering rocks do not cohere, and if the upper are lighter than the fluid and the lower heavier, then there will be a separation and the fluid will spread beneath those which are lighter and will not rise completely through them. If the density of the solid rocks increases downward more rapidly than does the density (under increasing pressure) of the fluid then the separation will take place at the point where the densities of the two are equal; and a very slight change in the fluid may give it a great change in choice of horizon for spreading. The case would be complicated by alternations of density in the solid strata. I think the separation would take place at a horizon

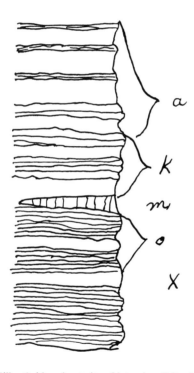

Figure 8.24. Gilbert's idea about site of intrusion (M) where dense beds (shaded) are overlain by less dense beds, p. 39 of Notebook 8.

{p. 39}   (*m* in Fig. 8.24) above which every combination of consecutive beds, including the lowest (as *a* or *k*) has a mean density less than the fluid, and below which every combination of consecutive beds (including the highest, as *o* or *x*) has a mean density greater than that of the fluid. Regarding the density of the fluid as a variable, the conditions give only one horizon for spreading to each value of the variable density. The removal of a

portion of the series *a* by denudation can have no effect in changing the horizon of injection unless the release from the pressure affects unequally the fluid and the solids, either directly as a matter of compression or indirectly as a matter of temperature.

Hence it is probable (but not certain) that the different levels of the Henry Mt. injections represent different densities

{p. 40}   of the injected lavas.

Did the Henry Mountains lavas reach the surface? All the direct evidence is negative, since no extrusive masses have been found, but their annihilation by erosion is conceivable. There could be no great difference between the density of a lava that would spread at the base of the Cretaceous and one that would rise above it, and the dikes which ran upward from it must (on hydrostatic rules) have gone very near the surface. If the lava contained a notable amount of water it would become inflated by steam at the top and lose density and overflow. An overflow once begun would hydrostatically be elected to the exclusion of injection and this would continue until an equilibrium was reached by the accumulation of a cone which the lava could no longer overflow. If the process of intrusion began it would be likely to open fissures for extrusion at a new point. In view of these speculations I am disposed to view the relation of intrusion and extrusion as one of antagonism and exclusion rather than transition and coexistence and to think that the Henry Mt. lavas were not extruded.

There are at least three

{p. 41}   conceivable conditions which might determine the difference between Henry and Uinkaret [*lava field, north of Grand Canyon*]. 1st. the solid strata at the surface and near it may have been less dense in H. than in U. 2d. the lava of H. may have been more dense. 3d. the lava of U. may have contained more water so that in rising to the surface it lost density more rapidly.

With regard to the first condition, the comparison is easy. The Uinkaret lavas rose above the Trias and the same lavas on the Upper Sevier rose above the Cretaceous and Tertiary. In the Henry Mts the highest known *sheets* are midway of the Cretaceous and many sheets could not pass the Trias, while some did not pass the Carb. We have reason to think that the Carb., Trias, and C. mean the same I think as regards density in each of the localities. Hence I think this condition was not the crucial one.

With regard to the second condition there is no direct evidence. It is impossible to restore deductively or experimentally, the conditions under which the Henry lavas congealed, conditions involving extreme pressures and temperatures. It is even impossible to examine

{p. 42}   dikes in which Uinkaret lavas have hardened under the same pressure. It is not to the purpose to determine the densities of the two lavas after congelation. Still an induction is possible from the coincidence of the Henry type of lava with the Henry type of structure. Within my own observation this is no exception to the law of coincidence, and the cumulative evidence of Henry, Navajo, Abajo, La Sal, Carrizo, Late (and one more making seven cases) goes far toward proving that the character of the lava affords the crucial condition. It may be the simple density as affected by water and inflation. Perhaps a critical comparison of Henry trap on the one hand with extrusive rocks and on the other with intrusive granite etc. may throw light on the matter.

It is a fact of great interest that faults are nearly or quite absent from the Henry Mts. Where strata are divided by dikes even no faulting appears. All displacements are by flexure

{p. 43}   and yet a displacement caused [*by*] the intrusion of molten lava cannot have been a matter of indefinite slowness, for there is evidence in contact phenomena that the molten lava was hotter than the solid strata, and prismatic structure makes a cooling of some rapidity. Certainly the mass of strata overlying a trap pocket was uplifted with a rapidity, under ordinary conditions would have separated it from its 'surroundings' by fracture and faults. What special condition prevented this? I think the condition of depth (resolvable into pressure and temperature) was sufficient. It is probable that these pockets were filled at depths ranging from 3,000 to 8,000 or more feet. The pressure would be from 300 to 800 atmospheres and the temperature due to depth would be (at a very low estimate) 90° to 140° Fahr. or (at a high estimate) 120° to 220°. I think the pressure would be more potent than the temperature in overcoming rigidity and preventing fracture—indeed I conceive it to be quite sufficient and a *vera causa*

{p. 44}   for the phenomena. With this view accords the contrast between the *ancient* folds of the Waterpocket and the Rock Hills regions and the 'post Tertiary' faults of the Sevier and Musinia region. [*Not post Tertiary—most of the faulting is middle and late Tertiary.*]

Is the Waterpocket fold a phenomenon closely akin to the Henry Mts and is it underlaid by trap? Possibly but not probably. The Waterpocket fold and its associated and details have the added features of lineation and *quasi* parallelism. The Henry Mts have no trend. Geographically they fall into line rudely but the general axis of the Pennell makes an angle of 45° with that of Ellen (if Ellen has any) and certainly with the trend of the range (calling the mountains for the moment a range); and the general axis of Hillers is at right angles to the range; and Ellsworth and H.V. have no notable major diameter. Perhaps I shall change my mind when I see the buttes on the flank of the Rock Hills

{p. 45}   of which Graves tells.

Friday, October 13. Bessie is a year old today.

Snow fell on the Mts last night while rain fell on our camp. The lowest snow is a thousand feet above us. Water in the frying pan indicates ¼ inch rainfall since yesterday. There was probably a half inch in the previous 24 hours. The wind (as shown by clouds) has been 'ESW' for two days and now it is nearly south. An eddy or return wind kept us yesterday at the verge of a line of cloud formation and 313 (200 feet above us) was half the time in cloud.

Sta. 26   8.57 am   22.40 = 8870 = pt 313 = an island of Belted hogback. At 12.50 = 22.38.

The Belted conglom. forms the whole footslope between here and the Hilloid. It dips about ten degrees and arches slightly downward toward the plain. It is nearly or quite unaltered and there is no trap with it until the birch spring sheet is reached. It gets nearly level at the base of 311 and has there a north rather than NW dip. 312*a* is trap but 312*b* and *c* are Belted. Southward the arch continues (Fig. 8.25) and the Belted forms 337 and the whole arched face A A. A A is a lower bed than 337 and 312*e*.

335—6-4 are bedded but are doubtfully trap capped. 338

{p. 48}   is on erosion slope and not structural.

In the cañon (339) the trap mass of the bulge is revealed and a dike shows at D (Fig. 8.26). It might be called the Bessie Creek Bulge or Lacune. [*This is the first time Gilbert used the term "Lacune." He had first written "Arch" but crossed that out.*] From this cañon W to the Gate Sandstone there is a continuous and harmonious dip that will average 15°.

Northward there is dip all the way to 327.

Camp 18 - 1.40 p.m. = 22.58.

Cache Camp 4.15 p.m. = 22.8500

Saturday, October 14. Cache Camp at 7:30 = 22.80 = 8380

Station 27 = 8:30 a.m. = 22.70 = 8500 = Point 240 S [*?*]. It is the south end of a grassy veneered ridge running north from Cache Creek near Cache Camp. At the south end the *"Gryphaea"* horizon is at the top of the ridge; at the north it is at the east base and the lower sand 'curves' to the top. Pt. 314 is near the north end (240 N [*?*]).

The west shoulder Lacune [*Pistol Ridge Laccolith*] (Fig. 8.27) is the largest single mass of trap that is visible about the mountains. There

{p. 51}   may be larger hidden. It can be traced continuously from Camp Graves to the base of Pt. 337 [Fig. 8.27]. All of the way it overlies the Belted and part way at least the *"Gryphaea."* Then the Belted can be traced continuously (p. 49) from its floor to the roof of the Bessie Creek Lacune which lies between the Belted and a lower bed that may be Vag. or Lower Belted.

The lower sand nowhere on this *coast* holds a hogback. It is too soft. Neither does the *"Gryphaea."* The weight of the different beds is about as shown in Figure 8.29.

Sta. 27 at 8:30   = 22.70 = 8500
            1:10 p = 22.74
            3   = 22.70
            4.40   = 22.72
Camp 1 at 5.10 = 22.82

Veneering. In the erosion of the Henry Mountains three zones are distinguishable. The summit, the flank, and the basal zones were originally distinguished by hardness and height—the summit being hard and high, the flank zone soft and high (but less high), and the basal, soft and low. In the summit zone erosion has been and is slow (relatively) on account of the hardness.

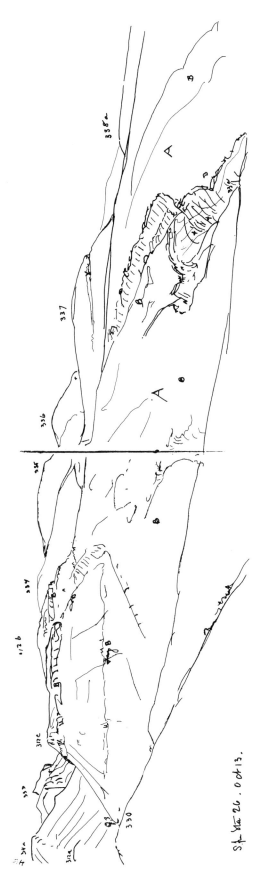

Figure 8.25. S from Sta. 26, October 13. The view is south along the distal edge of the Pistol Ridge laccolith (= Gilbert's Shoulder laccolite).

Figure 8.26. Another view of the west edge of the Pistol Ridge laccolith and arch over the underlying Cedar Creek laccolith. In his notes Gilbert referred to this intrusion as a "lacune," his first use of that term, and in his notes referred to it as the Bessie lacune or arch. In his report he named it the Shoulder laccolite.

178

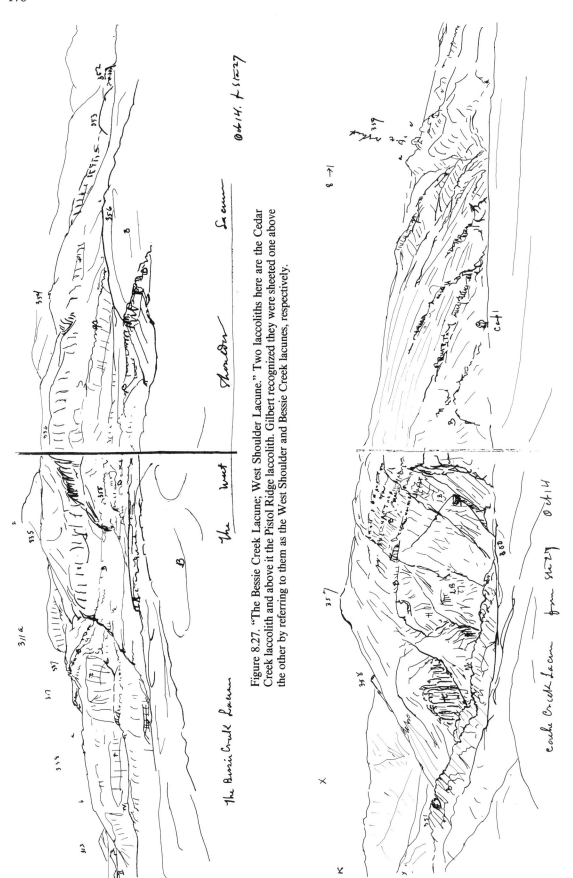

Figure 8.27. "The Bessie Creek Lacune; West Shoulder Lacune." Two laccoliths here are the Cedar Creek laccolith and above it the Pistol Ridge laccolith. Gilbert recognized they were sheeted one above the other by referring to them as the West Shoulder and Bessie Creek lacunes, respectively.

Figure 8.28. "Cache Creek Lacune, from Sta. 27, October 14." In his report, Gilbert referred to this as the Newberry Arch lacune (see his Fig. 39). It is intruded into Jurassic. In U.S.G.S. Prof. Paper 228, the intrusion is referred to as the Dugout Creek laccolith. High above it, possibly at point 337 in Gilbert's drawing, is a higher laccolith intruded into the Tununk Shale.

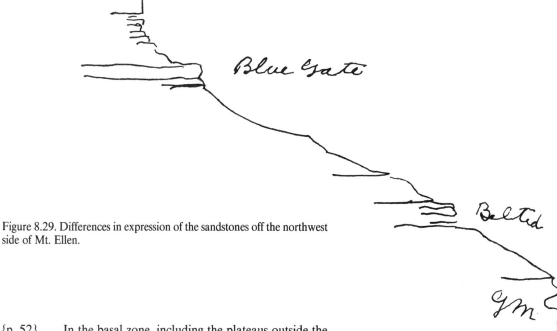

Figure 8.29. Differences in expression of the sandstones off the northwest side of Mt. Ellen.

{p. 52}   In the basal zone, including the plateaus outside the mts the erosion is slow on account of low grade. In the intermediate zone of soft but uplifted rocks, the initial corrasion was rapid and the declivity was quickly reduced to that which sufficed for the transportation of the products of summit erosion. It could go no further and lateral erosion began with the hard fragments from the summit zone as tools. Broad sloping deltoid plains have resulted—not strictly, for there was no filling, no hollows being possible when corrasion was simply working toward a limit. Great areas of soft rock were planed off and barely veneered with the debris of the summit zone. But the corrasion of the waterways of the summit and basal zones was not at an absolute standstill and as it progressed the head and base of the erosion of the flank regions were lowered and the veneering and planing process was carried on at successively lower levels. Remnants of early veneered plains were stranded here and there just as remnants of old flood plains often remain as terraces along river margins.

{p. 52a}   'Scheme'

An introductory chapter might describe the surroundings of the H. Mts., give their group character, altitudes, magnitudes, etc and tell how to find them.

    II. Ellsworth
   III. H.V.
   IV. Hillers
    V. Pennell
   VI. Ellen
  VII. The Lacune structure, this the largest and important chapter.
 VIII. The fold district
  IX. The fault district
   X. Erosion and sedimentation. Inconsequent drainage—veneering, cross-lamination, ripple marks.

Sunday, October 15, 1876. Rain all day. Cache Camp.

Boys returned from Rabbit Valley with supplies and letters. [*Mail service in Rabbit Valley?*]

Monday, October 16, 1876. Station 28 (ascended from Camp 1) = Pt. 27 = pt. 127.

9.4 a = 22.69 = 8500

10 a = 22.67

{p. 53}   The West Shoulder lacune may perhaps connect with a heap of trap across Cache Creek. The Mt. is quite tabular on top of it and it descends this 'name' of Shoulder. Here too there is coal on top of the Gate Sandstone.

340 is a veneer from the Cache Creek Cañon and the same may have extended over 317 and 318. The veneer on this butte is more likely from S. Cache.

The dip of the Gate mesas is very evident from this point of view.

Under 26b is a faint trace of bedding, nearly horizontal.

The West Shoulder appears to run across Bessie Creek and under 311a, 311 [*porphyry ridge, N side Cedar Creek*] and that spur to its junction with spur 2.

The rock of the upper course of S Cache Cr. seems soft and unaltered, but is not evenly veneered. I see no sandstone and assume it to be all Blue Gate Shale. I see no trap in place bet. 357 and 26a.

[[p. 54 is Fig. 8.30; p. 55 is Fig. 8.31. Pages 57 and 58 are tables of numbers apparently trying to calibrate aneroids 2 and 3 and are omitted}]

180

Figure 8.30. Unlabeled drawing on p. 50. Looks like badlands of Blue Gate Shale NW of Mt. Ellen.

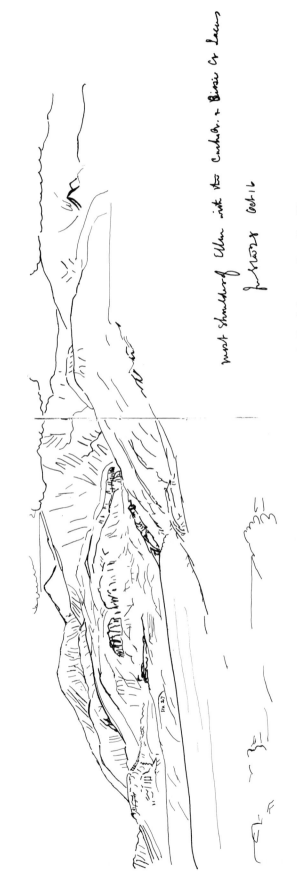

Figure 8.31. "West Shoulder of Ellen with the Cache Creek and Bessie Creek Lacunes, from Sta. 28, October 16."

{p. 59}   *Programme*

Oct. 9. March to 7 [*The Horn*]
Oct. 10. climb 7
Oct. 11. bench to Lower Cache and Slate
Oct. 12. Return to Camp 1.
Oct. 13. Pt 126. Rations
Oct. 14. March (Pt 26a) to Camp 16 = 114
Oct. 15. march and 223
Oct. 16. march to base Hillers

Oct. 17)
Oct. 18) S base Hillers
Oct. 19)

Oct. 20)
Oct. 21) Ellsworth
Oct. 22)
Oct. 23)

Oct. 24) H.V.
Oct. 25)

Oct. 26. march to Colorado
Oct. 27. station
Oct. 28. up Waterpocket

Oct. 29. 99 (Waterpocket)
Oct. 30. Adams pocket
Oct. 31. Bessie's spring
Nov. 1. Jerry
Nov. 2. 197
Nov. 3. 246
Nov. 4. March
Nov. 5. Pennell
Nov. 6. Summit camp and 22
Nov. 7. Ellen
Nov. 8. Cache camp
_____

Oct. 22. 223 and March
Oct. 23. march and cache for Ells.
Oct. 24, 25, 26, 27   Ells
Oct. 28, 29   HV
Oct. 30, 31   South Hillers
Nov. 1. Jerry and Pulpit
Nov. 2. 197
Nov. 3. 246
Nov. 4. march
Nov. 5. Pennell
Nov. 6. Summit camp and 22
Nov. 7. Ellen
Nov. 8. Cache camp

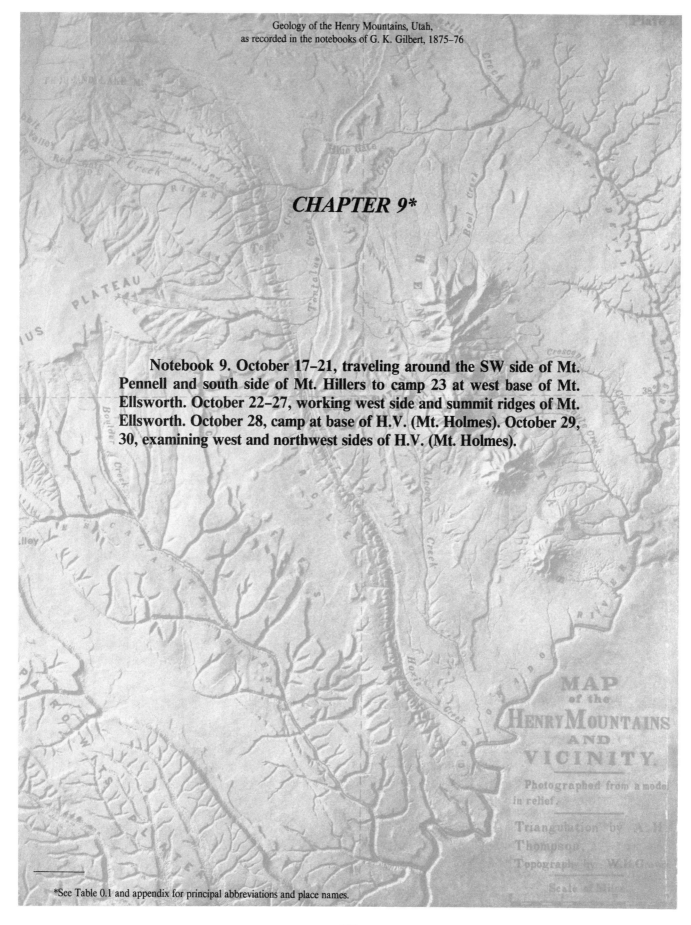

CHAPTER 9*

Notebook 9. October 17–21, traveling around the SW side of Mt. Pennell and south side of Mt. Hillers to camp 23 at west base of Mt. Ellsworth. October 22–27, working west side and summit ridges of Mt. Ellsworth. October 28, camp at base of H.V. (Mt. Holmes). October 29, 30, examining west and northwest sides of H.V. (Mt. Holmes).

*See Table 0.1 and appendix for principal abbreviations and place names.

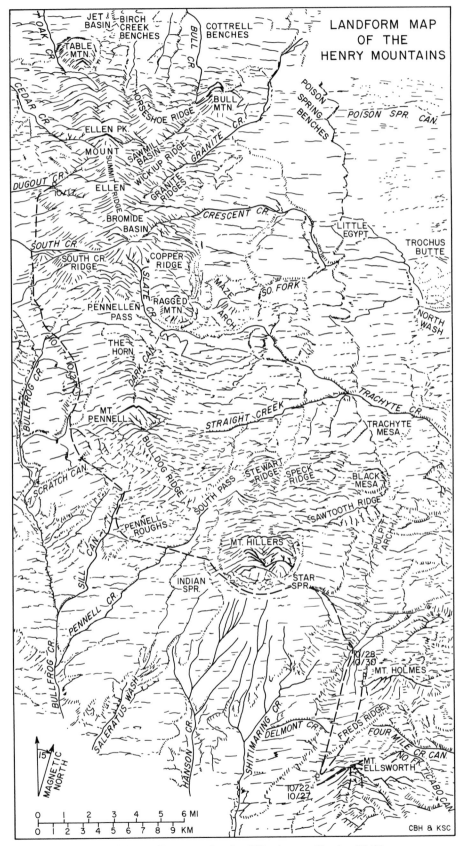

Figure 9.A. Landform map showing Gilbert's route October 17–30.

Figure 9.B. Oblique air photo of the west side of Mt. Pennell. The stock includes the peak and the summit ridge right of it. The west flank of the mountain is composed mostly of sills dipping steeply westward off the stock with the enclosing Cretaceous sandstones and shales. The prominent escarpment in the lower left is formed by the Emery Sandstone. At left center is the Horn laccolith; above it and beyond and slightly to the right is Ragged Mountain bysmalith. At right is the Hillers-Pennell Pass with Mt. Hillers in the distance. (Photograph by Fairchild Aerial Surveys.)

G. K. GILBERT, POWELL SURVEY, WASHINGTON, D.C.
NOTEBOOK NO. 9
OPENED AT CACHE CAMP, OCTOBER 17, 1876

{p. 1}   **Tuesday, October 17, 1876.** Camp 1 at 8 am = 22.68 = 8510

Valley north of the Gate Hogback at 8:30 = 23.03 = 8100

S Cache Creek at Pt. 341a (veneered annex) = 9.10 = 22.70 = 8500

Saddle E of 128 [*and at NW end of South Creek Ridge*] is 9.55 = 22.05 = 9290

The terrace noted yesterday under 26 a and b is made by a sheet of 100–200 ft thickness. At 21.55 = 9920 I am midway of its depth 'in fig'. I find however there is trap all the way up. At the first shoulder I read 10.30 = 21.28 = 10,240.

128 is evidently an old delta of S Cache Creek [*one of the oldest pediment gravels along the base of Mt. Ellen*].

The shoulder rises a little higher than noted and then is connected by a neck with the main spur. Two other fingers join it in the same way. In the neck is a narrow dike flanked by shale and in rising to the station I cross chiefly trap but some sandstone [*Ferron*] that probably has not great dip.

Sta. 29 = 26*a*. 11 am = 20.85 = 10,800. 12.35 pm = 20.88.

{p. 2}   There is sandstone outcrop farther up the hill—so scattered that nothing can be made of its dip except by regarding a number together. It dips 3°–5° to the SW and would project very little above the station.

The west face of South Ellen is very straight just under the crest line—even straighter than that of N Ellen. It is moreover very steep for 1,000 or 1,500 ft and then a quick change of grade and a shoulder. On South Ellen the shoulder has no abrupt edge and is the upper valley of S Cache Creek. But a trap sheet basis for the shoulder structure is revealed in many places. The axial ridge must be due to a dike for it shows no sheets and this spur (26) is dike-formed also. Then the Shoulder stands for a group of sheets and 'sheet'-Lacune. It is little better than a guess but it does not conflict with the few facts known to call the whole crest of Ellen dike-born.

Camp 19 is a half mile upstream from camp 10. At 6.45 = 23.09.

Starting to descend from the station I find shale below the trap (or at its side). When near camp a back-sight at South Ellen showed conspicuously the straightness of the west side.

{p. 3}   **Wednesday, October 18.**
Camp 19. 6.15 am = 23.03 = 8100
          8.6   am = 23.04 = 8090
Camp 16. 8.24 = 23.10 = 8010
On the bench (above Gate sand) above dinner camp bet. 15 and 16 = 10.4 = 23.42 = 7650.
   Dinner camp bet. 15 and 16 = 10.15 = 23.37 = 7470
   Pine Alcove under pt 138 [*or 38*] at 12 n = 24.12 = 6940.
   We leave Pine Alcove near the last W outlier of the Gate 50 ft above creek. I read 1.15 pm = 24.22 = 6720.
   Bench 24.15 = 6800. 135-136 wash. 1.40 = 24.30 = 6640. Bench again 24.20 = 6740.
   Yesterday on Sta. 29 we had a few snow flakes, white frost, and ice in canteens.
   A terrace (noted from sta. 11) is very conspicuous on the NW flank of Hillers. It is above the line of springs and, considering its height, is too definite and persistent to be a veneering. Still there are veneering lines below it.
   Dinah Cr. at 3 pm = 24.15 = camp 20 (Fig. 9.1).
   Camp 20 is on Dinah Creek not far below Dinah Ranch. [*Dinah Ranch is probably a spring-fed grassy oasis on the gravel veneered benches sloping west of south from the Hillers-Pennell pass. It is named for a horse left there.*] It is in badlands below the Gate sandstone. The badlands are far removed by long straight washes which acquired their straightness by sliding down smooth conic (almost plane) veneered plains. The process can be seen in its stages. The badland ridges between the washes are straight only in a general way and their summits are level only in a general way (Fig. 9.2).
   From each node there runs a lateral

Figure 9.1. An unlabeled drawing on p. 4; probably badland ridges.

{p. 5}   spur either to the right or to the left. The nodes at the right hand spurs are to the right of the median line as a rule. I wonder why. It is evident that the sending of a spur should determine a node, for it diminishes the mean declivity of the slopes draining that point of the creek. But why should erosion be more rapid opposite a spur?

These forms are the result of erosion under remarkably uniform conditions of hardness and original slope and their study should reveal laws of erosion.

One law seems evident—that declivity diminishes as quantity of water increases. The arroyo has less declivity than the surface which feeds it and the great arroyo than any of its branches. But there is a limit to this on the crests of the ridges. The increase law of declivities and quantities of water would make them sharp edged and in fact they are rounded. Is the condition rain-beating?

Figure 9.2. "The badland ridges between the washes are straight and their summits are level only in a general way, as varied as in the diagram [*evidently a profile*] but with smoother curves."

{p. 6}   Is it disintegration as distinguished from transportation?
Thursday, October 19.
Yesterday or rather last night was the most notable exception in regard to wind. It blew up hill or rather with the strike all the evening and after the shower was quite gusty. This morning however it blew down hill as usual. In camps 17, 18, and 19 followed the usual rule.
   Camp 20.   7.25 am = 24.21 = 6730.
   The Gate shale is so full of the thick *"Inoceramus"* and they are so divided that no square foot of the surface is without the fragments. They abound in the gravel of the creek bottom.
   We start toward 273 climbing a hill of slidden trap and sandstone fragments on shale. The shale seems to be chiefly in place. At 23.56 = 7480 we are a little below the sandstone of 135*a* (500 ft below its summit). This at 9 am. At 9.25

{p. 7}   I have risen to 22.98 = 8150. I am now on the level of 135*a* in which are developed some minor flexures. The rock below me is the lower sand with the coal on top of it. It dips 15° or 20° from 134. Above it to the left is a narrow long sheet of trap following the slope and conforming.
   Rising higher we leave the lower sand again and travel on Gate shale.
   Then we come to a sheet once more. It may be the same one, but it is thicker. Its horizon seems to be toward the base of the Gate shale. It is one of the set which finds its culmination in

Figure 9.3. View north along the beds and trap sheets turned up along the SW side of Mt. Pennell.

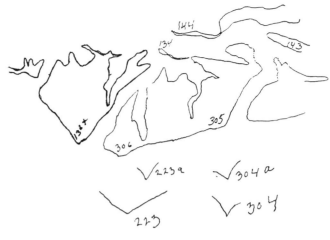

Figure 9.4. Outline of the Gate sandstone as sketched by Gilbert on p. 10 of his notebook. North is to the right.

223. It has a rise of 15° or 20° where we climb on it, and the dip increases to 30° toward the mountain. The dip of 223 is 32°–34°—almost the stable angle. The trap between 223 and the mountain has a sheety facies parallel to 223 [*SW flank of Mt. Pennell; hill 9345?*] and the strike of the great spur which runs from 74 [*The Horn*] to 134 is parallel to the sheets.
11.23 = 22.53 = 9930

Camp 21 is at the base of pt. 223,

{p. 9}   a little stream that may not be permanent rises and sinks among some aspen and pine. At 1:30 pm 21.24 = 9680. **Friday, October 20, 1876.** Camp 21 at 6:30 = 21.46 = 10030. There was a snow flurry last night. ¼ inch of snow fell.

Wind. Yesterday pm the wind was south or largely up the cañon. About dark it changed and blew down the cañon all evening. It still blows down.

Sta. 30 = pt 223 =	8.25 a	20.41 = 11,400	
	11.00 a	20.33 = 11,500	
	11.00 a at fire,	20.38 = 11,430	
	11.45 a at fire,	20.38	

The arroyo on which we are camped [*Mud Creek*] runs between 135 and 136. North of it the Little Gate sand [*Ferron Sandstone*] rises in a point only twice as distinct as 223a and with the same bearing as 135d [*Emery Sandstone escarpment SW flank of Pennell*]. 306 connects with 223a. Between it and the point first mentioned is a cañon which joins Pine Alcove opposite 144 [*canyon next southeast of Scratch Canyon*]. Another heads against its S wall halfway down. A through cañon heads at 305 and joins P.A. [*Pine Alcove*] Creek at 143. [*This is Scratch Canyon.*]

{p. 10}   It is some miles to the next one, which passes the 'old' dinner camp.

The higher Gate salient is 306. The one S of it is about as high.

Work almost barred by a snowstorm and gale of wind.

Sketch on this page gives the outline of the Gate (Fig. 9.4).

Camp 21 again. 2.35 pm = 2151 = 9970.

Descending I noticed slate chips with the trap, but I see no slate masses. At several

{p. 11}   points are masses of trap different from the prevalent which is fine-grained and similar to that of Sta. 27. I think the coarse is lower in the bedding.

I think our camp has the altitude of 306. Level frozen. Theodolite levelled by guess and the horizon.

Camp 22 is on upper Dinah Creek on the base of Hillers. At 7.15 pm = 22.27 = 9000.

Leaving camp 21 we returned on our trail of yesterday to the point from which I sketched 223 and then continued parallel to the 134 crest finding very good travelling on a broad rolling bench of which the rocks were trap, sandstone, and shale not often seen in place and all probably slidden to some extent. We crossed a stream (21.73 = 9680) a mile before we came to the cliff at the pass (Hill-Pen). The cliff is not a clean wall as it is east of the Pass and at Sta. 11 but is a heap of slidden masses of sandstone. I was surprised to cross some trap that seemed in place during the descent.

{p. 12}   **Saturday, October 21**

The trap which I crossed last night in the dark I can now see in the continuous belt running WNW from Hill-Pen pass. It appears to dip NE but there is so much sliding that I do not feel sure.

The Hillers shoulder above the spring line is a trap sheet and the debris includes sandstone and shale. [*Gilbert refers to the porphyry sill in the Tununk Shale south from Cass Creek to near Squaw Spring.*]

I can now see the folding of 135 island [*No Man Mesa*] and that it is simple (thus, Fig. 9.5) with a NW strike.

Later I find a sheet below the Hillers shoulder and travel on it for some distance. The section at one (numbered) point is Figure 9.6.

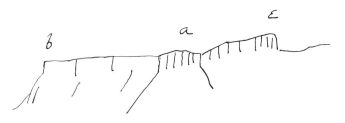

Figure 9.5. No Man Mesa from Stanton Pass.

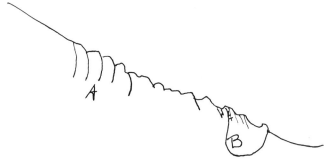

Figure 9.8. South slope of Mt. Hillers showing the zone of trap (A) and the J-Triassic turned up at the south base of the mountains.

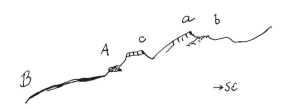

Figure 9.6. Gilbert's cross section of the west base of Mt. Hillers. B is the "sheet we have crossed and A the one under the shoulder." The "numbered point" referred to in the text (p. 13) is not shown in the sketch.

{p. 13}   Halting just after joining the trail by which we left Hillers before, I can see beyond this section a promontory of sandstone rising toward the Mt. (Fig. 9.7).

I think it must be the lower sand. It is too thick for the Gate and between it and 135 [*No Man Mesa*] there are almost continuous outcrops of shale.

These data lead to an addition to my idea of the general structure of the Mt. for these dipping trap sheets overlie and conform with a great trap sheet [*Stewart Ridge laccolith*] which forms the spur N of the west cañon of Hillers and are thus connected with the northward dip of the Belted on 66. [*Farmer's Knob?*] The most westerly point of that spur is a half mile beyond the general base of the Mt. and consists of a lone outcrop of lower sandstone [*Ferron Sandstone*] striking about due north and dipping 10° to the west. From the mouth of the cañon southward the dip is exhibited by uprising sheets, and swings

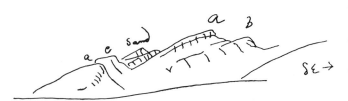

Figure 9.7. Promontory of sandstone northeast of Stanton Pass. It probably is the Ferron Sandstone dipping northwest off the Stewart Ridge laccolith.

{p. 14}   rapidly around the base to the sandstone buttresses. I read at the end of the sandstone spur, 10 am = 22.78 = 8400.

The south spur of Hillers [*the stock, which forms the main high part of the mountain*] is an immense mass of trap of which there are tremendous exposures in the region A (Fig. 9.8). The region B shows only J-Trias. The Lacune must be supra-Carboniferous. At the extreme south there seems to be an abrupt change of strike, and the dip almost reaches 90°. I am writing where the lower sand cliff rims the Mt. by running under the veneering.

Saddle bet. veneered hill and the mouth of the South Cañon of Hillers at 11.40 = 17.88 = 8280, a little 'apron' of sandstone dipping 10° from the Mt.

Beyond the cañon the first buttress shows almost no Trias but a nearly continuous exposure of trap. Part of the trap is dike and this is flanked and finished by sheety masses bet. which are occasional glimpses of

{p. 15}   sed.

The line of max flexure makes a sharp salient at the mouth of the cañon. For a distance of 1–2 miles to the west it is straight (Fig. 9.9).

Dinner camp 20 ft lower than Camp 14. 2.5 pm = 23.38 = 7680. We cache here some flour, Jerk, corn beef, sugar, beans, coffee, yeast powders, milk, paper, rocks, and a box.

Incidents. Yesterday Joel rolled over once on a hill and smashed my looking glass. Today the cook's fire ran through the oak leaves and burned my saddle, saddle blanket, spurs, overcoat, and saddle bags—all a little so that repairs are needed.

Top of Belted Cliff 4.10 = 23.98 = 700 [*7000?*]. Pass at foot of cliff 24.11 = 6840.

{p. 16}   We begin to rise a veneered branch at 24.15 = 6810 and reach the top at 4.32 p = 24.03 = 6930.

Then into a cañon in veneered Vag. 24.10 = 6860 and out again 24.00 = 6970 = 4.48 pm.

Figure 9.C. Oblique air photo of the east and south sides of Mt. Hillers. The steeply dipping sandstone formations are turned up around the stock and its shatter zone which cut discordantly from Permian wall rock at the south base to Cretaceous wall rock around the north side. The easily eroded Upper Jurassic and Cretaceous formations form the gravel covered pediments sloping from the mountain. Pulpit Arch is at the right center and beyond it is Sawtooth Ridge (Gilbert's Jerry). Mt. Pennell is seen beyond Mt. Hillers. At the right is Mt. Ellen. On the skyline (left) is the Aquarius Plateau with the hogback of the Waterpocket Fold (white) along the near side of that mountainous plateau. (Photograph by Fairchild Aerial Surveys.)

Camp 23, at the base of Ellsworth at 7.30 pm, 23.25 = 7840 at the head of the veneered bench.

Sunday, October 22. Camp 23, at 7.45 = 23.26 = 7836
Mouth of cañon = level of camp 23a = 8 am = 23.26 = 7830

At 8.55 = 22.02 = 8650 we are on a level with point on the next great spur west of this one. [*Point identified only as a.*] It is a trap sheet (?) above the purple streak of the G.M. The upper crag is lower

{p. 17}   in horizon but still above (just above) the purple band. They may both belong to a dike.

The spur we climb is a sheet superior to the Aubrey and in

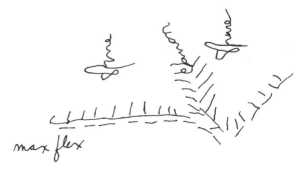

Figure 9.9. Map showing salient in the line of maximum flexure at mouth of canyon at the southeast corner of Mt. Hillers.

its debris are perfect feldspar crystals which I gather. [*In some intrusions the feldspars break down by weathering to pits, whereas the hornblende phenocrysts weather free, and intact crystals can be collected. In other intrusions the hornblendes break down, and the feldspars weather as intact crystals that can be collected. In still others the paste survives both kinds of phenocryst, as described by Gilbert, p. 16, Notebook 4. No explanation for this contrast in weathering has yet been discovered.*]

At 10.13 = 21.96 = 9400 I am on the saddle above the great knob of this spur and have trap all around me. In the cañon at the west I see purple and red rocks as well as much trap but do not make out the structure except that there is a dip arching from the core.

$$Sta\ 31 = pt\ 220 = 10.40\ am = 21.88 = 9740$$
$$1\quad pm = 21.65 = 9790$$
$$4.30\ pm = 21.61$$

There is an exposure of purple under the south spur with 500 ft of trap over it. Surely an immense depth must be found to assure one that he has the lacune of such a mt. as Ellsworth.

Monday, October 23.

Camp 23   7 am = 23.28 = 7800 is just outside our camp of last year being above the Shinarump instead of below it. It is on the variegated marls here colored chocolate, red, and slate. Dip 46°.

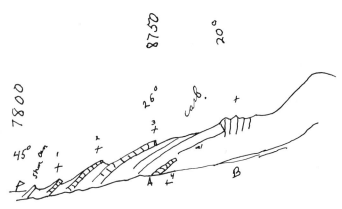

Figure 9.10. Section exposed in canyon wall above camp 23; sketched by Gilbert Monday, October 23. Station numbers referred to in the notebook pages 17–18 are not shown in the section. Location of crystals is shown.

{p. 18}   The first sheet is 50 ft below the Shinarump conglomerate and a dike just west of the cañon connects it with the second sheet which is within the chocolate shale. The shales are almost unaltered but are a little harder than usually.

Some of the trap surfaces show streakiness and I find a specimen in which this extends ½ inch into the body and then *gradually* diminishes, the long white streaks changing to feldspar xx [*crystals*]—as though the xx had been pasty.

The third sheet is 70 ft thick (2d—20—50; 1st 15) and rests

on the Aubrey. At 8:10 = 22.57 = 8660 I read its dip across the cañon 26°.

It is the third sheet that yielded the xx yesterday.

The fourth sheet shows at bottom of the cañon opposite (A. p. 19) and just above it there is a purple streak within the Aubrey. Then the Aubrey ends abruptly against a great body of trap to which I see no further limit. Its base however is covered by talus and it may have a sed. floor under the talus. It has not however the prismatic cleavage of the sheets. I do not see the floor of the fourth sheet, nor definitely its base. It surely underlies 200–300 ft of Aubrey quartzite (for it is altered more than the Shinarump group). At the head of the Aubrey I measure dip (20°) and altitude (the base is about on my level = 22.41 = 8840). I am 75 ft above the top where I read 9.5 = 22.16 = 7130 (pt. 393b).

{p. 19}   The mass B has a counterpart in the climbing spur with two knobs 396a and 396b. b is 30 ft lower than a. On a level with a I read 9.20 = 21.90 = 9480.

Bedded in its flank is a mass of green slate which I am unable to refer. So far as quality goes it might be Cretaceous but it is more likely to belong below Trias.

Sta 31 reading immediately on arrival

9.30	= 21.70 = 9720 (an shaken)
9.31	= 21.70 (shaken)
9.50	= 21.71
12 n	= 21.70
1.30 pm	= 21.69

Sta. 19 is not a dike as I thought, i.e., the trap does not penetrate the rock below. Is it a lacune? or a lacune-dike? I think the sed. under it is Vag. Certainly that of 216 is a lower bed. But the separations between lower Vag and Gray Cliff and between Gray and Verm.

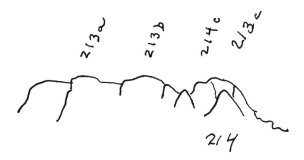

Figure 9.11. "Halfway (or less) up the slope I am able to pick out the true 214" (p. 21).

{p. 21}   are so similar and there is so little to distinguish the sandstone masses that I am in doubt as to the horizon of many masses.

Spur 275 shows purple beds (probably of Shinarump group) in its flank near its summit (200 ft below 384).

191

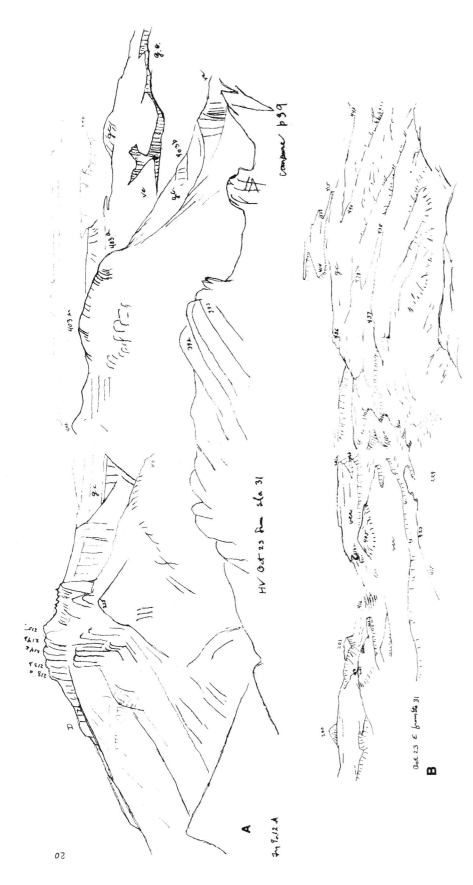

Figure 9.12. View north and east from the north spur on Mt. Ellsworth, Gilbert's Sta. 31. A, view north to Mt. Holmes (cf. Fig. 9.11); B, east to the mesas and canyons bordering the Colorado River. These are not satisfactory drawings; difficulty was experienced identifying features on the microfiche. These illustrations are included chiefly for the sake of completeness, to illustrate that Gilbert made notes about the surrounding country from every vantage point. The north spur on Mt. Ellsworth is one of the difficult places to reach. I was there with plane table and alidade. As Gilbert wrote (1877B, p. 119), "One may ride to the crest of Mt. Ellen and to the summit of Mount Pennell; he may lead his sure-footed cayuse to the top of Mount Hillers; but Mounts Ellsworth and Holmes are not to be scaled by horses. The mountaineer must climb to reach their summits and, for part of the way, use hands as well as feet." My guess is that Gilbert was tired, as I was, when he reached Sta. 31.

In A, the central mass of Mt. Holmes probably is the west side of the Theater laccolith against a background of dikes at the stock and extending eastward and westward from it. In B, the broad, short canyon at the left probably is Two Mile Canyon. The canyon from the lower left and away from the observer at the center (along the crease between two of Gilbert's pages) probably is Four Mile Creek Canyon. And the canyon crossing the lower right probably is the North Fork of Ticabo Creek Canyon.

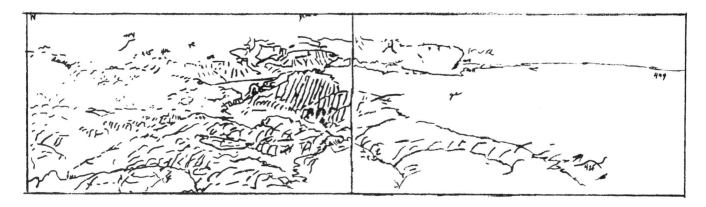

Figure 9.13. October 23. Southeast from Sta. 21.

Tuesday, October 24. Camp 23 = 7.20 = 23.27 = 7810. Last year's camp 7.30 = 23.25

To the E of camp there is a sheet resting on Shin. Conglom. and it visibly thins downward.

Sheet 2 is composed of 2 parts. Halfway (or less) up the slope I am able to pick out the true 214 (Fig. 9.11). A mistake was made day before yesterday.

Note. A specimen collected yesterday was incorrectly labelled sheet 4. It is from the general mass with which sheet 3 merges above. Sta. 31 = Ellsworth. At   8.57 = 21.78 = 9620
                                                            11.45 = 21.76

Saddle between 220 and 384 = 21.90
384 = 21.81 = 9590
Sta. 32 = pt 385
12 n = 21.85 = 9630

From sta. 32 it is plain that the great mass of trap we avoid in climbing is a dike. It does not reach sta. 32 in 396a. It is joined by another great and equal dike which runs to the bottom of the next cañon. Others reticulate with them as in the diagram (Fig. 9.14) and within the reticulations are sedimentaries (aaa p. 25) [Fig. 9-15]. At the base *BBB* are sheets dipping not less than 30° between Shinarump beds.

This spur is held partway by a dike but shows chiefly sandstone of cream color and purple shale. They dip 10° to 15° toward 160 (NW?) but toward 220

{p. 25}   dip diminishes. The dip in 275 which looks more like massive sandstone than trap is westward and 20°. If it is Trias it is altered. I think sta. 32 and its spur must be low Triassic.

Yea verily 275 is massive sandstone and so is a great mass of the 385 spur. I see not even a dike in 275 until the saddle 275-385 is reached. The rocks arch over toward the spur and the max. dip is at the base of the mountain.

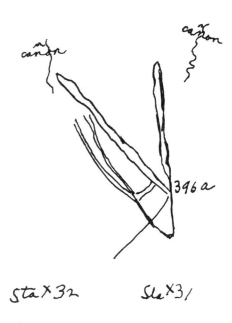

Figure 9.14. Plan of dikes seen from Sta. 32.

{p. 26}   Spur 388 [*west spur of Mt. Ellsworth*] is built chiefly of Gray Cliff sandstone sustained by a reticulation of dikes, not sheets, nor do I recognize a sheet in the west cañon. The end of the spur seems to contain no trap, but the sandstone has a very different facies from the same bed beyond the base. It is full of seams and rusty.

The spur cannot be sketched to advantage but the diagram on this page (Fig. 9.16) gives the structure. I think nothing below Vermilion is exposed but if that is true, then the purple belt is not here hardened to sandstone as in 275 and at camp 23.

Figure 9.15. North spur of Ellsworth from Sta. 32. Reproduced as Figure 11 in Gilbert's Henry Mountain report.

Figure 9.16. Sketch of the West Spur of Ellsworth.

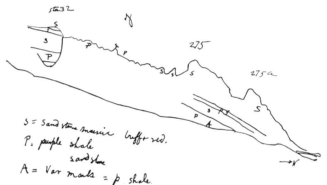

Figure 9.17. Face of spur 275, North Spur of Mt. Ellsworth.

Descending the 275 spur I find a dike in it but the entire mass of 275 and 275a is sandstone of the G.M.

{p. 27}   All goes to confirm the interpretation given to the spur and the dike from the station. The sandstone is somewhat shot with xx but is still a sandstone in texture rather than a quartzite—light, open and granular.

The lowest point on our journey to camp, bet. spurs in the upper Shinarump reads 23.40 = 7640. Camp 23 at 7 pm = 23.25 = 7840.

The sun set at 5.25. It will rise 'horizon' at 6.30.

Wednesday, October 25. Camp 23   7am 23.20 = 7900

The base of the Vermilion has the finest exhibition of mud-cracks I have ever seen. They reticulate with interspaces of five feet or more and their casts are 5 to 10 inches through at top. I note one which penetrates the shale for ten feet.

Our camp creek enters Verm. at 7.30 = 23.47 = 7580.

455b (= 7.43 = 23.34 = 7720) is also at the base of the

Vermilion which dips about 40° and is steadily curving in strike around the mountain base.

Looking across 455 cañon near the foot I can see 275 well. The Verm. makes the long spur which runs to 455b. The var. in mts show below it, but the Shin. Conglom. is covered. There is a 100 foot

{p. 28}   purple band above the Verm (if it is above *all* the Verm. then the latter holds not over 250 ft) and above that 1,000 ft of Gray Cliff making 275 and 275a.

500 ft below Sta. 32 is a purple mass showing 250 ft (and see p. 30). It is probably the Verm and in that case the purple *on* 32 is the purple band or something just above it which I do not see on the 275 face. It is the same horizon which gave me purple shale on the high saddle (sta. 32 = 275) last night. There is notable dip under Sta. 32, say 15°. I do not think the dip under 275a exceeds 35° (Fig. 9.17).

{p. 29}   *455a* = 8.20 = 23.00 = 8130 at this point. A sheet
stands at the east of the cañon with a body of the Chocolate Shin
below it. It is about 75 ft thick (with annexes outside) dips 45°
and strikes toward sta. 19. It must be below the Shin. Conglom.
At 22.76 = 8410 its highest point as a buttress. Just where it joins
the cañon, it joins also the great dikes of the cañon. In the opposite wall of the cañon is a sliver of sandstone included in trap. It
shows a max. thickness of 10 ft and is 25 ft in height. Its dip is
from the Mt. and it may be in place. Collected from its contact.

Between the two great dikes (A p. 25) [*Fig. 9-15*] I see a
smooth broad exposure of trap which looks as though it was sheet
or lacune. The rock exposure above it is purple (and green?) and
the dike (?) above that has a definite flat upper limit as though it
had terminated under a hard stratum. This same evenness of
upper limit can be seen too (obscurely) on the great dikes. I
suspect it too on the summit of H.V. [*Mt. Holmes*].

Purple shale, under the sandstone, in saddle (32 and 275
[*North Spur*]) (y. p. 28) 9.15 = 22.17 = 9110. This shale and that
on 32 do not reduce in

{p. 30}   weathering to an even clay like that on the buttresses,
but become coarsely granular with chips ¼ to 1 inch across. In the
cañon below, the pockets and interstices between rocks are filled
with this purple shingle instead of the usual sand.

At 22.10 = 9200 I am on a level with the trap of the purple
under 32 (p. 28) and can see that the dip there is deceptive. The
strike and not the dip is toward Sta. 31 and the dip is 20° to 25°
toward 455b. The thicknesses are far less than I estimated—on
the nose of spur 32 however the dip is 20° toward the west base
of 275.

I think I can now see a mass of trap on the saddle adjacent
to 275 and another on the continuation of the same bedded in its
west or SW flank. If they conjoin they are a dike.

        Sta. 32   9.45 am = 21.80 = 9600 (an shaken)
                      9.46   = 21.775   (shaken)
        Sta. 32a   9.47   = 21.77
                      10.21 = 21.77
        Sta. 32   10.22 = 21.75

Under sta. 31 is a body of white and ochreous, 'curious' rock
(SW flank) which seems to merge into trap at the south. I take it
to be a decomposed trap. [*Porphyry in parts of the stock and
shatter zone are altered hydrothermally to white porous rock in
which constituent minerals are not obvious.*]

        Sta. 33   11.40 am = 21.05 = 9530
                      3.20 pm = 21.73 = 9700

{p. 31}   From Sta. 32 the S face of spur 285 is seen. It is
traversed by a fine reticulation of vertical dikes (Fig. 9.18). The
ground plan I have sketched shows sufficiently their general character and magnitude. There are no sheets discernible.

Between these dikes is a great expanse of the sedimentaries
which do not seem to have a consistent structure under 32. All is
Gray Cliff for 500 ft (except trap). Then there is a little show of

purple shale. Then below the dike *aa* there is another exposure of
sandstone and another of purple shale, this time clearly the variegated. These show some small faults. Beyond the dike *cc* there is
still another mass of sandstone with shale on top and then more
sand and beyond the dike *cd* and the cañon is a series consistent
to the Buttress 9463. I have sketched in diagram the section
(p. 32, Fig. 9.19) and indicated by *xx* the best correlation I can
make. Certainly there is faulting here in connection with dikes.

PS. I see one small sheet at A p. 32.

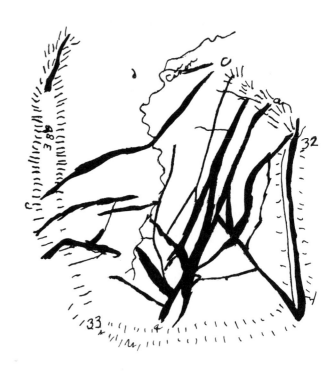

Figure 9.18. Reticulated dikes in the face of 285, seen from Sta. 32.

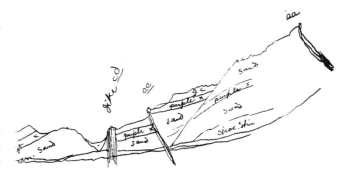

Figure 9.19. Faulting associated with dikes on northwest side of Mt.
Ellsworth, under Gilbert's Sta. 32.

{p. 32}   Looking across the south cañon at the south spur—the section at the end is plain. A dike crosses the spur obliquely near the end and while the block west of it conforms with the buttress, that at the east is lifted 200 ft or more. The sketch next page shows all this and gives only the *end* of the spur (Fig. 9.20). The body of it shows many dikes and a great deal of debris that must be sedimentary but no strata in place.

Sta. 33 and nearly the whole crest of the spur is of sandstone.

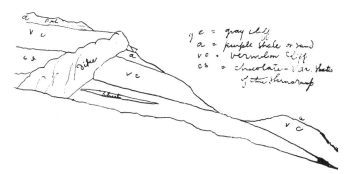

Figure 9.20. Sketch of the end of the spur off the northwest side of Mt. Ellsworth. Notebook 9, p. 33.

{p. 34}   Near 31 I find the wind ripples strongly marked. The waves are 9 inches from crest to crest. The material is an angular gravel of disintegrating trap. The mean size on the ridges of the particles is 3/16 inch, or 2 'lines', the max is 1 inch. The interspaces are filled with fragments from ½ to 3 inches in size. The ridges slope downhill and there is no soil among the pebbles of the interspaces, think it is washed out. The pebbles of the ridges are bedded in earth, or at least their interstices are filled.

The curious rock noted on p. 30 *is* decomposing trap and it looks very much like some metaliferous bodies. It is the best place I have seen to look for a mine. [*Gilbert is on the Mt. Ellsworth stock. On the northern mountains the stocks are much larger, and they are higher and more watered as Gilbert described. Each of the northern stocks is cut by a major canyon with sides mantled by colluvium; the shady south slopes are forested. This cover obscures features distinguishing the stocks from the laccoliths— their steep side walls discordantly cutting across the sedimentary formations, their greater shattering, differentiation, metamorphism, and teasing signs of metallization.*] Close by and in similar decomposed condition is a conglomerate or breccia of sandstone in trap and the same occurs again beyond sta. 31.

Sta. 31 at 4 pm = 21.55 = 9900. Saddle at the east 21.75 = 9670.

Sta. 34 that is to be at 4.30 = 21.75 = 9700.

{p. 35}   **Thursday, October 26**
Camp 23.   7.20 a.m. = 23.11 = 8,000

The spur we climb this morning is the one we came down last night and leads directly from Sta. 34 to camp 23. 20 ft below the crest of the first trap sheet I read 7.31 = 22.94 = 8210. At 7.46 (= 22.69 = 8500) we are 25 ft lower than the crest of the second sheet. So far as I can see eastward the flank is sheeted to this height and 100–200 ft higher. The third sheet veins far up this spur but just east of the trail it joins a dike which cuts off the view.

At 8.25. 22.21 = 9080.

I am sitting on a high point (not quite the highest of sheet 3) and looking east. The sheet continues beyond the first cañon but with diminished thickness and it does not climb so high. In the cañon there is intimation of a lower sheet that must divide the Aubrey. The sheets numbered 1 and 2 appear in the next spur but not beyond. The sheet which in that ridge first appears above the Shin (No. 0) holds for two ridges more and is then replaced en echelon by one just about on the Shin. The diagram will help show the

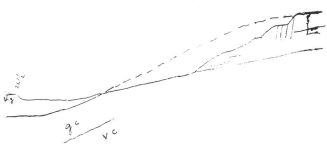

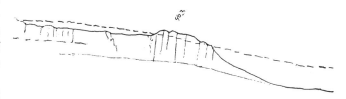

Figure 9.21. "This is my idea of the structure of H.V. The dikes are probably given too small. They can be corrected by the mapwork altitudes (compare p. 20)." From Notebook 9, p. 39.

{p. 36}   overlapping although it is very rude in proportions.

Sheet 3 seems to rest directly on Aubrey.

I have no hesitation in saying that Sta 19 is on Variegated (*Chinle Formation*] and pt 216 on Gray Cliff. The purple band at the top of the Vermilion has a soft streak over it which undermines the Gray Cliff. The whole top of Little Henry is Gray Cliff. The Vermilion runs well up the south cañon and runs just under a sheet that starts there. Think there is a thinner sheet lower down but can't see color. But with these exceptions the whole face of H.V. is dike—sustained.

The ridge which divides camp cañon carries Aubrey as far

as I thought before and probably a little further, but it is cut off this side same as the other.

{p. 37}   Sheet 3 on this spur joins dikes and its identity is lost.

Where I come upon Aubrey (opposite its crest in next ridge) it dips only 10°–15°. A little farther I find a dip of 45° toward 59.

Farther up the spur it diminishes but at the farthest exposure it still has a dip of 20° from Sta. 31. I trace the Aubrey (the top of it I think) to within 100 yards of Sta. 34 and then all is trap.

Sta 34.   9.28 am = 21.72 = 9700
          11.15 am = 21.74
          11.45 am = 21.78
          2.70 pm = 21.69

The proximal end of the south spur looks from this side just as from the other—a mass of dikes containing perhaps sed. and (less likely) perhaps decomposed trap. The spur which runs east from the station has dikes in the nearer portions, and at least some sheets in its remote.

The limits of the Ellsworth disturbance are hard to determine from Ells. I can trace dips S to 443 and almost to 446.

Sta 19 [*Fred's Ridge*] is on the neutral ground between Ells. and H.V. and a little N of the structural saddle.

The H.V.-Hillers divide (structural) is in the neighborhood of 269½

{p. 38}   H.V. dip are traceable plainly to 420 and less surely to 206a. Trachyte Cañon at its mouth makes about the limit.

South of the mountains is a great system of joints. They are revealed wherever the Gray or Vermil. Cliff rocks weather in slopes, but are not equally pronounced at all points. They are sub-parallel and separated by irregular intervals (Fig. 9.D). Adjacent joints do not vary more than 1° or 2° from parallelism and none vary much from verticality. The general trend is SE. I can see the joints 20 miles away to the south. If they exist in other directions they cannot be seen because viewed from the wrong angle.

My measurements of dip today from Sta. 34 are probably all too small (at least for Ells. buttresses) since the same method gives 399 too small.

Descending S to a crag—

In the spur leading to 223 [*this number duplicated*] I can discriminate no sed. clear down to the saddle inside the buttress (base of Var. shales). Sheets perhaps exist but they are inconspicuous. The chief exposures are dikes. The next spur N has a sheet in the Var. shales—higher up, dikes.

In my topography I failed to get to the summit of Buttress E.

{p. 40}   Between Buttress B and 199 are only two (Fig. 9.22).

The spur M is heavily sheeted below the Shinarump but above a thinner sand of the chocolate. There is no Aubrey east of N nor west of the climbing spur.
Section

Sta. 36 is 34 ft of strata above the base of the purple

I measure 132 ft horizontal for the base of the Vermilion to

pt. 492. The Var. marls occupy this ground and extend to the Shinarump.

Friday, October 27. Camp 23. 7 am = 23.17 = 7920

I think I have solved the problem of the rounding of the crests of the badland ridges.

If a solid of wedge form be moved into a space of a higher temperature, it will gradually heat by conduction from the surface. From this surface *AC* (Fig. 9.23) inward to *ac* the heating will have progressed equally and a gradation of temperature produced and in like manner from *AB* to *bd*. But the area *Aaeb* being acted on from both sides will be

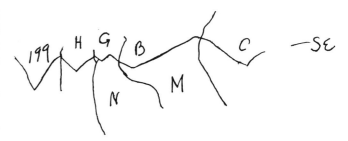

Figure 9.22. "Summit [*peaks?*] between 199 and Buttress B."

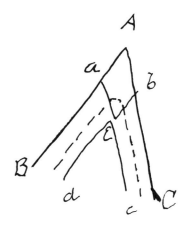

Figure 9.23. "Penetration of temperature [*or of moisture or frost*] is greater where the surface is convex than where it is plane, enabling angular corners to become rounds. Likewise the crests of badland ridges."

{p. 41}   more heated than if acted on from one side only and a line of equal temperature run midway between BA, AC, and *deec* will be curved about *e* instead of angular.

The same principle applies to any process which penetrates from the surface of a solid. *When the surface is convex there is greater penetration than when it is plane and when it is concave there is less.*

There is a good illustration in the curves produced in the

Figure 9.D. Vertical air photograph showing jointing in the canyon-forming sandstones referred to by Gilbert on p. 38 of Notebook 9 (compare Fig. 4.D). Arrow shows approximate north. Scale about 2 inches equal one mile. Ticabo at mouth of Four Mile Creek canyon is at lower right. Mouth of Red Canyon entering the Colorado River from the east is a mile north of that. This is, of course, pre–Lake Powell. (Photo by Fairchild Aerial Surveys.)

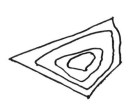

Figure 9.24. Lines of equal weathering in the fragment of angular rock, illustrating the increasing rounding inward.

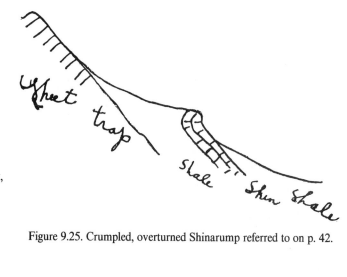

Figure 9.25. Crumpled, overturned Shinarump referred to on p. 42.

weathering of an angular block of stone. Crack one open and you find the lines of equal action curving and with a gradual decrease inward of resemblance to the outward of the rock (Fig. 9.24).

It is under this law that angular blocks of crystalline rocks (or other) become round by simple weathering, without attrition.

Now what is the action which penetrates in the case of the badlands? It is weathering. It is the action of water in dissolving, in decomposing, in expanding (either with or without frost) the shale and thus disintegrating it and preparing it for transportation.

Thus in the badlands, transportation and corrasion are tending to produce angular forms, and weathering

{p. 42}   opposes their production. The action of weathering is most potent when the deviation from a plane is greatest. Angularity is greatest where erosion is most rapid.

From 399 I can see Elisha's pasture. For ¼ mile there is no notable drainage across the Shin. and it stands as a retaining wall for the shale above and trap debris (Fig. 9.25). For the whole distance the upper edge of it is broken and crowded over, a phenomenon that does not occur where it is cut through by drainage lines and rises in buttresses.

I now think that the purple under Sta. 32 is upper Shin. and that on its summit is supra-Vermilion. Then the dike south of the crest marks a displacement.

The sheet which crosses or ends in Cañon 455 is like No. 3 of the climbing ridge and No. 2 runs almost as far. No. 1 reaches only 20 rods from camp. [*Refers to Fig. 9-21A?*]

{p. 43}   The section series of stations will give thicknesses for the Shinarump to the base of the Vagabond. The following description will complement it.

At the base of the Vagabond series—
1. Deep red sandy shale   50 ft
2. Massive sand, banded with red and pale red   40 ft
3. Massive pale-red sand   500 ft
    (2 and 3 make the Gray Cliff)
4. Purple and deep red sandstone (bedded)

5. Red massive sandstone (4 and 5 make the Vermilion and its parting)
6. Shale, chocolate above and variegated with purple slate color below (Shin., wood)   200 ft
7. The Shin conglom   20 ft
8. Chocolate shales with some aren. and calc bands   350 ft
9. Aubrey Sandstone

The *Vag Series* is exceedingly difficult to estimate as it makes broad valleys and is veneered. My impression is that on the scale of the above section it makes 1,000 ft.

My measurements of dip seem wild. I think 45°–50° about the general dip of this section.

{p. 44}   *Mountain sculpture and climate*

I have before written that the orders of size and ruggedness of the mountains are reciprocal. The following is its tabular expression.

N = Ellen, P = Pennell, H = Hillers, E = Ellsworth, V = H.V.

1 Altitude	2 Area	3 1 and 2 combined	4 Smoothness
N	N	N	N
P			P
		P	
H	P H	H	H
	E		
		E	E
E			
	V	V	V
V			

There is no difference in the constitution of the mts to account for this phenomenon; it is purely a matter of climate, as I conceive the mts which lift the largest masses highest get most moisture. This stimulates vegetation and thereby aids disintegration (weathering) and hinders transportation. The result is a smoothness of slope and contour as compared with lower smaller mts.

In general, the other conditions of erosion being the same, aridity tends to ruggedness, to cañons, to pinnacles.

Figure 9.E. Oblique air photo looking east across Mt. Holmes (Gilbert's H.V.). The mountain is in the canyons eroded into the Navajo and other Jurassic and Triassic sandstones that have been domed by the stock at the center and high part of the mountain. The stock is small and uplift is less than at the other stocks centering at the laccolithic mountains (see Fig. 0.B). Numerous dikes, sills, and small laccoliths radiate from the stock. (Photograph by Fairchild Aerial Surveys.)

{p. 45}   Left camp 23 at 1 pm. At 2.15 the Hill., Ells., saddle 24.06 = 6910. We follow down the wash to 3.15 = 24.35 = 6580.

Searching for water I come upon a magnificent section in a cañon [*modern Cache Creek*] of H.V., which leads toward Jerry. A sheet sends a salient down on the west side of the cañon and my station is 200 ft below that salient. I sit on Vermilion and it extends 100 ft below me, but there is a fault along the cañon and on the wall the base of the Verm. is 100 ft above me (23.32 = 7760) (Fig. 9.26).

Figure 9.26 is the cross section and Figure 9.27 is the opposite wall.

Camp 24 = 5.40 pm = 23.62 = 7400

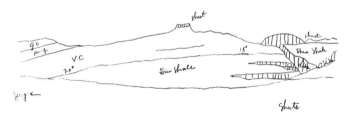

Figure 9.26. Cross section of Cache Creek fork of Star Creek. Compare Figure 9.28.

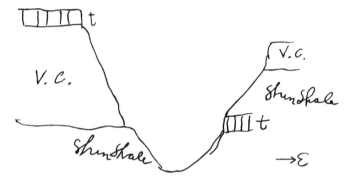

Figure 9.27. Wall opposite that of Figure 9.26.

{p. 47}   **Saturday, October 28.** Camp 24 at the foot of H.V. [Fig. 9.E] 7 am = 23.54 = 7500

Yesterday in approaching the mt. I thought I saw several distinct great blocks on the west side at different levels. This morning in the NW spur I see a dike and a great sheet. The dike is obliquely longitudinal crossing to the west in descending. The sheet shows below so that the section is like Figure 9.28.

Leave horses at 8.10 = 23.23 = 7860

Our camp is at the base of the purple band.

The trap sheet described above has a very decided diminution in thickness. At the upper end where it joins the dike it is not less than 200 ft thick and the bottom is covered by talus. At the lower end where the bottom is visible it is 50 ft thick and there is an outlying

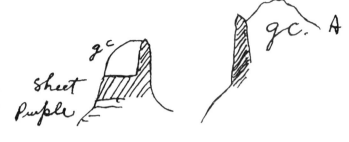

Figure 9.28. Sections across the intrusions forming the west spur of H.V.

{p. 48}   fragment still thinner. The dip of the top is 20°–25°, thus (Fig. 9.29).

The dike east of the ascent slopes noted yesterday has a dip of 25°–30° and above it is another with no more than 10°. It too (z, in Fig. 9.30) has some taper. At 22.68 = 8820 I am 100 ft below the top of the upper.

The body of sandstone (A, Fig. 9.28) is crossed also by the great dike (D, Fig. 9.12A). The point at which they join reads 22.35 = 8900. Seen from this side the sandstone makes a shoulder of which I have measured the height (Fig. 9.31).

The lower dike trends toward pt. 379 (*"Gryphaea"*).

There is a sheet below the sheet Z (supra) and which

{p. 49}   I noted last night buried in the Verm. sand.

At 9.39 = 21.93 = 9450 I am in a cleft at the east of 213, a cleft 200 ft deep that is probably not a cañon by fissure.

The great NW sheet is parallel to the one of which the direction is noted above. It has a sheer face extending 50 ft lower than this point.

Sta. 39a = pt 213 or 213b at 10.8 = 21.71 = 9720

Sta. 39 at 3.15 pm = 21.62 = 9230

Sunday, October 29.

The sheet past which I climbed yesterday is not the one I had noted the day before but is of a higher horizon and seen only on the low spur toward camp. It rests on the lowest part of the purple band above the Vermil. Between it and the other sheet, which also is tongue-like, is a little dike that has turned adjacent sandstone white.

The sandstone between the great NE and NW dikes

{p. 50}   extends within 200 ft of the top.

Near the base of the top crag of H.V. are a few Douglass spruces and with them are large piñons which I mistake for yellow pines. [*So did I.*]

Camp 24. 7.45 am = 23.44 = 7500.

End of dike on camp spur = 8.17 = 23.07 = 8040. I get on the spur at 22.20 = 9090.

The Island Butte of trap which I sketched (p. 46, Fig. 9.27) has a counterpart on the spur beyond. I will call the first 497 butte and the second 498, and use the same numbers on the spurs.

Figure 9.29. Thinning of trap sheet (p. 47–48).

Figure 9.32. "Pulpit Arch, as seen from the flank of H.V., October 29, 1876."

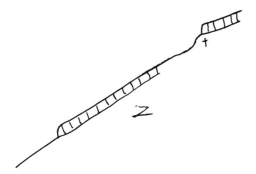

Figure 9.30. Thinning of sheet referred to on p. 48.

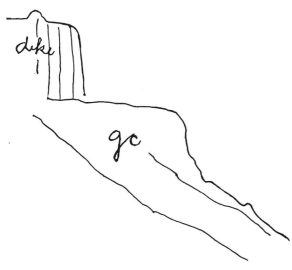

Figure 9.31. Shoulder formed by G.C. sandstone where crossed by the great dike in the west spur of H.V.

497 is sheeted above the fault in the sketch and carries the Shin. Shale nearly to this point. Sheet 498 is sheeted as far. The fault crosses it also. It is broader and thick. Its dip is about 20° at the butte and less above. A dike runs along on top of it and has an even top parallel to the sheet as though ending against a strata. I see no Vermilion in either space above the fault.

{p. 51}   Sta 39.   9.45 = 21.70 = 9730
                    12 m = 21.64
                    2.20 = 21.65
          Sta. 40.   2.50 = 21.90 = 9470

{p. 52}   Sta. 40 is at the head of spur 498. The remote sheet of this spur is above a position at least of the Vermilion. Beneath the nearer Shin. Shale shows. There is a still lower sheet toward (toward 504) that is apparently traceable to the base of 403 (which by the way is the same as 217) and its more remote portion appears to be concave upward.

The whole ridge to 403 seems dike-borne though holding much sandstone.

There certainly is a sheet and a great one running down the cañon south from 213. It joins the Mt. 1,000 ft below the summit and is 300 ft thick, tapering to 200 [*Theater Canyon laccolith*].

The Var. shale under 503 are white and there is little Verm. on them. It has been cut off by veneering.

{p. 53}   (Figure 9.33)

{p. 54}   I think I sighted on pt. 5 Saturday instead of Ellen. To test it solve the following.

```
I = Ellen  P = Pennel  R = pt. 159  E = Ellsworth
     H = H.V.  T = Two 46
RIP = 180 - PRI - RPI
RIH = RIP ≠ PIH
RHI = 180 = RIH - IRH
THI = THR - RHI
THI = THR - 180 ≠ IRH ≠ PIH ≠ 180 - PRI - RPI
     THI = THR ≠ IRH ≠ PIH - PRI - RPI

    THR        IRH        PIH        PRI        RPI
  281.38      75.43      57.59      25.5      357.46
  194.59                 34.54                229.32
   86.39                 23.5                 128.14

     86°39'             22°5'
     76     43          128    14
     23      5          153    19
    186     27
    153     19
   33°8'  = THI computed

         103°53'
          70   3
          33°50'   THI observed October 29

         191°25'
         158  21
          33° 4'   THI observed October 28
```

Figure 9.33. Summit of H.V. (Mt. Holmes) from Sta. 40.

From this it would appear that the first identification of Ellen was correct.

Monday, October 30. Camp 24   7.55 am = 23.60 = 7440

Left horses in pothole gulch (23.90 = 7080) and climbed to sta. 41 = pt. 271 = 9 am = 23.65 = 7380.

This butte is a veneered island of Gray Cliff. The Gray Cliff sandstone is here as red as the Vermilion; in this morning light it looks even darker than the Vermilion at camp.

{p. 55}   The line of max. dip is a little outside this point.

Badlands. The occurrence of a groin opposite to a spur (p. 5) and the consequent zigzagging of the crest of a badland ridge is on the principles of the honeycomb. The arroyos *eat back* and when those of opposite sides meet—1st, that their line of demarcation is straight (supposing their progress to be equally rapid) and in the direction of the common tangent of the two courses of radiant "eating"; 2nd, that the point of this line which is first defined by the touching of the curves is eaten most rapidly.

If the arroyos *a, b, c, d, e, f, g* (Fig. 9.34) meet in the plain *p* and cut it all away. The crest line *FG* will have apices at *A, E, B, C, D* and from these points will run off a side spur, a spur for each salient of the zigzag.

Wind. At camp 24 has been S each night, gentle and shifting during the day.

{p. 56}   (Figure 9.35)

{p. 57}   Horses under Sta. 42 at 2.40 = 22.65 = 8550 (see p. 52).

Sta. 42 at 3 pm = 22.22 = 9070. Sta. 42 = pt. 258. At 4.10 = 22.19

The eastward buttresses of Hillers (259, 260, 373) are all of trap and in line. The two former rest on beds of Vag. sand and outside there are less inclined Belted Con. beds. There is an interhogback or interbuttress valley along the line of the Vag. and

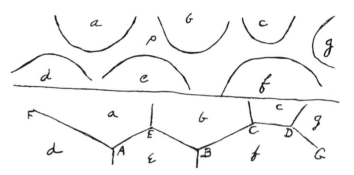

Figure 9.34. Gilbert's geometric explanation for the zigzagging of the crests of badland ridges (p. 55 of Notebook 9).

then comes the G.C. hard and pinnacles. Beyond it trap in chief part. There are dikes, chiefly narrow running out to the capsheets of 260 and one comes between 257 and 258 (Sta. 42). Westward in 257 there are sheets in Belted and the G.C. runs into the Mt. being greatly masked by trap. Mountainward dikes multiply until there seems to be trap only. [*This is the poorly exposed cross-cutting east wall of the Mt. Hillers stock.*]

The sheet islands of H.V. are explained by the fault. On the Mt. sed. which was uplifted, they (the sheets) were carried too high to survive erosion, but outside just a vestige remains: 497a, 271a, 498a, and probably 498a are at the same horizon and doubtless one sheet. The H.V. disturbance extends to 500 and almost to Trachyte Creek. In that structure are two arches (Fig. 9.36). Possibly the change is along the fault line.

Camp 14, 5.45 = 23.15 = 7950 (Fig. 9.37).

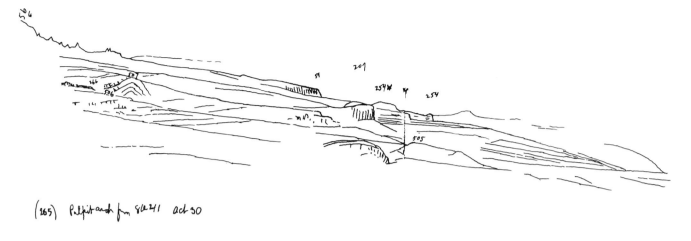

(265) Pulpit arch from Sta 41 act 30

Figure 9.35. Pulpit Arch, from Sta. 41, October 30.

Figure 9.36. Two arches interrupt the north flank of H.V. (p. 57, Notebook 9).

East Buttresses of Hillers from Sta 42 Oct 31

Figure 9.37. East buttresses of Hillers from Sta. 42.

CHAPTER 10

Notebook 10, opened at Camp 14, near Starr Spring, October 31, 1876. Traveling north along the old trail, about at the position of the present highway to Camp 25 on Trachyte Creek. November 1, explores Crescent Creek and Trochus Butte; finds Indian tracks in trail to North Wash. November 3, camp in aspens at east foot of Mt. Pennell, a previously occupied camp with names cut into trees (p. 15). November 4, notes missing. November 5, cross Penellen Pass to head of Bullfrog and then to South Creek and from there to base of Hilloid Butte (= Table Mountain; camp at a spring along Oak Creek?). November 6, climbs Mt. Ellen. November 7, "Homeward bound. Election day." November 8 in badlands as far north as Factory Butte (Gilbert's Needle Butte). Records crossing Indian trail twice yesterday. November 9, crossing Belted Cliff (= Morrison Formation). Dikes and sills in Vag. (= Entrada Sandstone) and Belted (= Morrison Formation). November 10 and 11, travelling NW toward Ivie Creek. November 12, "Elisha kills a deer at Dinner Camp." November 13, reached the Cretaceous shale, *"Gryphaea"* bed near Ivie Creek. November 14 along Ivie Creek crossing Castle Valley. November 15, reach Gilson's Ranch at fork of Ivie Creek. November 16, obtained horse from Gilson; travel to Salina Canyon. November 17 and 18 to Salina and Gunnison. November 19 and 20 at Gunnison. November 21, Gunnison to Nephi. November 22, Nephi to Salt Lake City (by railroad).

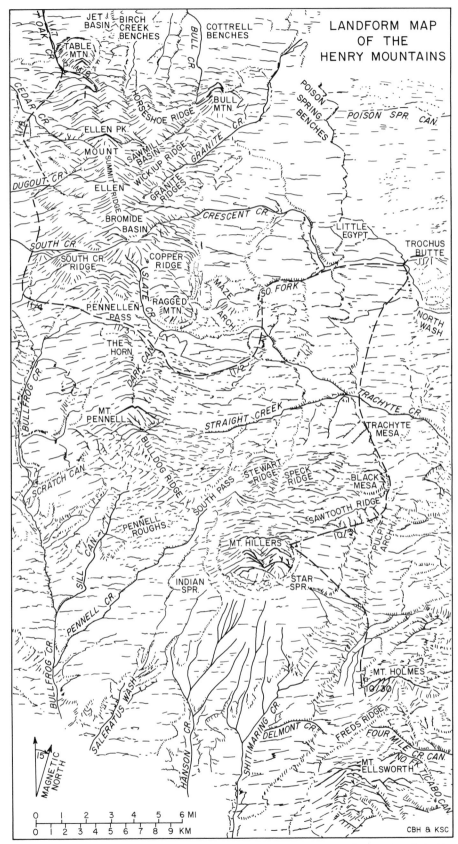

Figure 10.A. Landform map showing Gilbert's route October 31 to November 7 where he left the Henry Mountains area en route to Ivie Creek and from there back to Salt Lake City (November 15).

Figure 10.B. Oblique air photo of the badlands crossed by Gilbert on his way north from the Henry Mountains, November 7 and 8, 1876. The mesa in the distance is Factory Butte (Gilbert's Needle Butte).

G. K. GILBERT, POWELL SURVEY, P.O. BOX 806,
    WASHINGTON, D.C.
OPENED AT CAMP 14, NEAR STARR SPRING,
    OCTOBER 31, 1876.
RETURN TO SALT LAKE CITY, NOVEMBER 22. 41 P.

{p. 1}   **Tuesday, October 31, 1876.** Camp 14 at 8.45 am = 23.25 = 7830; a touch of snow last night. Last year's campground at the head of the water 8.53 = 23.10 = 8,000.

Sta. 42 is upon a rock which exhibits the rounding of boulders *in place* with particular beauty.

All the trap in H.V. is coarse grained.

My 'perch' for sketching today is at the base of 257 [*east base of Mt. Hillers*] in line with camp 14 and the right flank of 215. Last year's camp projects to the right of 216.

The water from the great south cañon of Hillers is surely that which rises at the camping ground but no surface water runs that way. It looks as though surface water from the cañon sank just outside its mouth and dropped its load. That being the case there can be no change until the coarse debris in which the water sinks shall have become built up with fines. If the interstices of gravels were not filled with fine material there would be little surface drainage anywhere.

257 is a sheet of trap above the Belted Conglomerate. At top its dip is 45° and it is 50–75 ft thick. Lower it can be traced continuously to the mouth

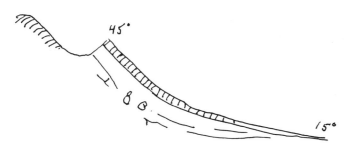

Figure 10.1. Sheet of trap above the Belted Conglomerate referred to on pages 1 and 2.

{p. 2}   of the adjacent cañon where its dip is less than 20° (say 15°) and its thickness is 10 ft. Here is a second instance of a trap sheet concave outward (PS. It is a dike). Its blocks are very angular in contrast with those of 258 (= sta. 42). I must sample it for comparison (Fig. 10.1).

The hornblende which I have before noted as occurring in blocks in the trap has gradually been explained. It is the altered condition of small fragments of shale etc. caught in the lava. Larger masses show less alteration and a transition. In one instance I found a lump with a core of hornblende and garnets outside. [*We first supposed this to be so, but later investigation*

showed that the amphibolites and shale xenoliths are not gradational. See U.S. Geological Survey Professional Paper 228, p. 160–164. See also U.S. Geological Survey Professional Paper 294-I, p. 349–352.]

Badlands. The penetration curves mentioned in Notebook 9 are further illustrated and in a most *a propos* manner by the rounding of boulders of homogeneous rock with attrition. The weathering of granite in rounded bosses—of this trap of Sta. 40, of the trachyte of Thousand Lake Mt and of the massive Triassic sandstones are instances. The area of points described for Ellsworth shows a great deal of this work (p. 6) (Fig. 10.2, 10.3, 10.4).

{p. 6}   (fm p. 2) The massive homogeneous sandstones (Gray and Verm) are divided by joints, and in these, water finds free access so that they as well as surface of exposure form bases for penetration phenomena. The frost (and perhaps other disintegrating agents) acts most powerfully and rapidly in the corners, and rounded bosses are the results—a wilderness of domes.

The penetration principle ought to be illustrated and demonstrated experimentally and demonstrated *a priori.*

An experiment is easy. Take a prism of a homogeneous conducting substance of which an end shall have some such shape as Figure 10.5

{p. 7}   *A priori* the temperature of an atom is influenced by (in conduction) the temperature of adjacent atoms in all directions, and the rate at which it changes depends on the difference between its temperature and the mean of all those temperatures. It heats more rapidly if heat comes from two sides than if it comes from one only.

When a profile of a homogeneous mass of rock is in the form of Figure 10.6, a, the slope *a* degrades most rapidly. In *b* of the figure, the degradation still is greater at *a* than at *b,* and so it is in Figure 10.6, c.

{p. 8}   But how can a divide move when every rain that falls on it must be split and sent in two ways? The explanation lies in the principle of penetration, in accordance with which a knife edge divide cannot stand and the steeper declivity practically saps the divide (even in homogeneous material).

The protection given to shale by veneer is 'feeble', but I doubt whether a mesa of pure clay would hold its mesa form during denudation.

At 2 pm = 23.65 = 7380 I am on the bluff near 506. The edge of the Pulpit Arch is at 505, and is abrupt, a change from 0° to 10° taking place in 100–200 ft. I am 50 ft below the level of the Cliff Springs. Cliff Creek must carry water a mile and may take it all the way to Trachyte Creek.

Camp 25, on Trachyte Creek below Camp 13. About 50 ft below top of G.C.

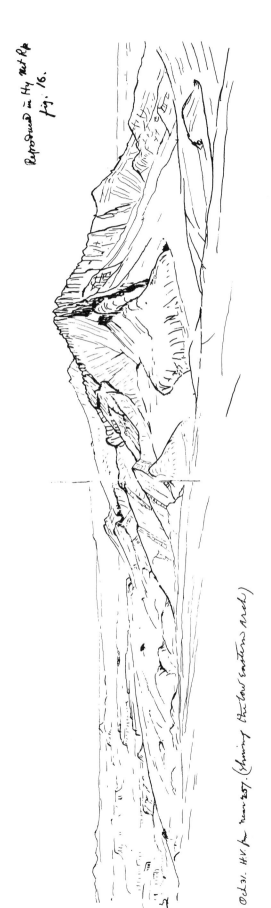

Figure 10.2. H.V. from near 257 (showing the low eastern arch). Reproduced in Gilbert's Henry Mountains report as Figure 16.

Figure 10.3. South Mountains, October 31. "Ellsworth should be tall as H.V."

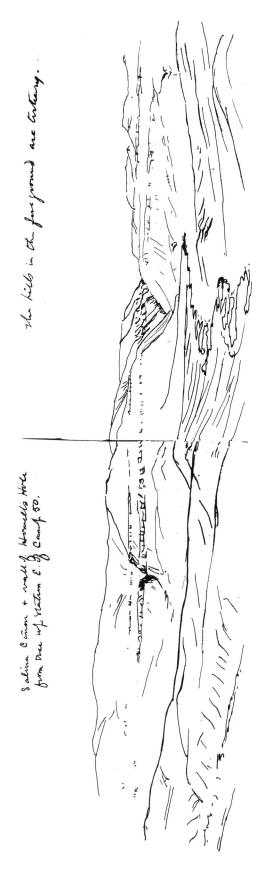

Salina Cañon + wall of Howell's Hole
from tree up station E. of Camp 50.

the hills in the foreground are Tertiary.

Figure 10.4. "Salina Canyon and wall of Howell's Hole from tree-top station E of camp 50. The Hills in the foreground are Tertiary." (I am puzzled why this drawing is so far forward in this notebook; it would seem to belong near the end of the book. Perhaps two pages were inadvertently skipped and filled in later.)

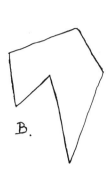

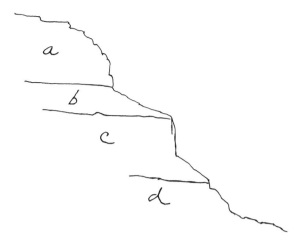

Figure 10.5. To show the principle of penetration phenomena, Gilbert suggested taking a prism of homogeneous material having shape A or B. "Have the surface of the end smooth and coat it with wax or lard thinly and evenly. Immerse it except for the end in hot water and watch the progress of melting of the wax; then in cold water and watch the freezing. To assist investigation and make it quantitative, draw lines parallel to the periphery of the prism and note the time they are reached in different parts. To sustain my hypothesis the progress of heating should be more rapid from the convex surface than from the concave, and the time of equal action should curve about all the reentrant 'angles' and across all salients." Notebook 10, p. 6 and 7.

Figure 10.6. Section of the Belted Cliff between Trachyte and Crescent Creeks.
"*A* White conglomerate and purple shale
b slate cold shale sloping
c chocolate or purple and slate cold. Shale or badland rock, more or less fluted (more at the N.)
d red badland sandstone—the Red Vag."

{p. 9}　　　5.15 pm = 24.54.

Wednesday, November 1. Camp 25 = 7.45 = 24.65 = 6250

Gray Cliff Sandstone is crossed by Crescent Creek at 24.71 = 6160

Horses 10:50 am = 24.11 = 6840.

Sta. 43 = 246 Trochus Butte. 11　　am = 23.98 = 6990
　　　　　　　　　　　　　　　12.45 pm = 23.94

Trochus point is high up in the red Vagabond. All along the Belted Cliff the section is like Figure 10.6

{p. 10}　　　The Belted Cliff is very level.

The Maze Arch, the Crescent, and the NE annex are distinctive but adjoined. Sta. 4 lacune [*Copper Ridge laccolith*] rests partly in the sag between the Maze and the Crescent. The NE Lacune may rest on the crops between Crescent, Annex, and Bowl Creek. The Sta. 3 lacune [*Wickiup Ridge*] is a true one but only half exhibited by erosion. The Pennell Arch seems more perfectly coalescent with the Maze than any other pair.

I can now see a tapering sheet stretching from the base of the NE butte NE over the N end of the annex. It seems to underlie and overlie Belted conglom [*can be seen in Fig. 6.D*]. The NE Butte has a half scaly or buttressed aspect about the outside edge.

We find today a trail made by one horse, shod or partly shod, and then horses barefooted, barefooted mules, and barefooted colts, in all about 15 animals. There is a mocassin track with them. They came down Crescent Creek, started up the trail toward Trochus Butte and stopped; one went ahead and turned

back and then all went down Crescent Creek Cañon. After an interval, they returned and went back up the creek again. The coming tracks were made in wet sand; the going in dry. Neither have been rained on. The tracks are much scattered.

Our last storm was October 20. From all this we infer that a party of Indians not familiar

{p. 12}　　　with the country came down Crescent Creek Oct. 21 or 22 and after an interval of some days (long enough to go to the Colorado and back) returned. They were less numerous than their (15) animals. It is not unlikely that they were Navajoes who had stolen stock from a stock range and were trying to cross the Colorado without passing through the settlements.

Thursday, November 2. Camp 25. 8.33 am = 24.67 = 6220

Top of G.C. = 24. 63

Camp 13 = 8.48 am = 24.60.

Trachyte Creek at top of red Vag. Near Sta. 14. 9.55 am = 24.32 = 6620

{p. 13}　　　Base of conglomerate = 24.21

These two observations are within ½ mile and give the thickness of intervening shale.

Water runs in both branches of Trachyte Creek at their junction. The strong and worked trail up the bluff between them shows that the cañons are not passable through the Belted.

But when at 11.42 = 23.98 = 7000 I have reached the top of the conglomerate I have lost considerable by dip and am only as high as the bench where we left horses on Sta. 14.

212

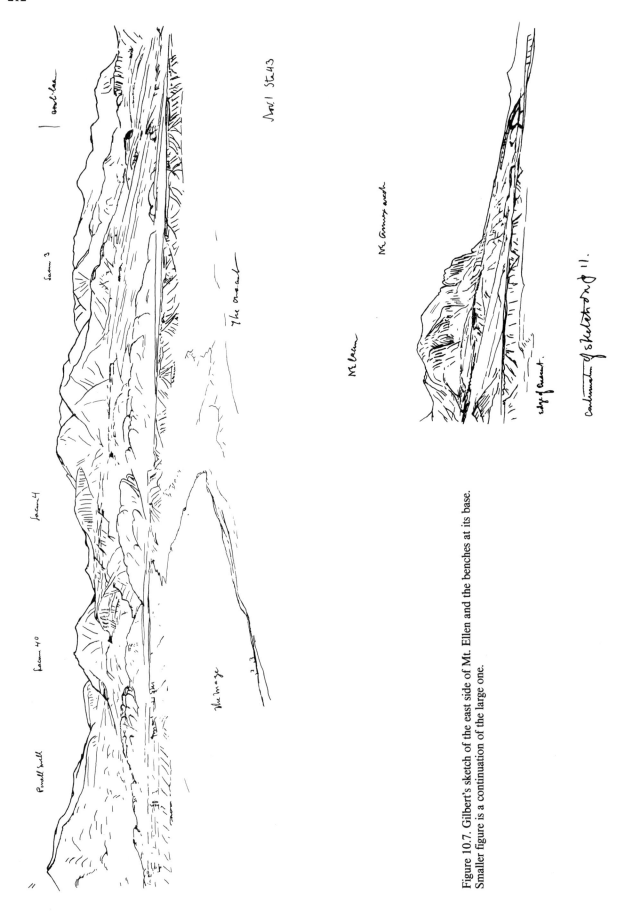

Figure 10.7. Gilbert's sketch of the east side of Mt. Ellen and the benches at its base. Smaller figure is a continuation of the large one.

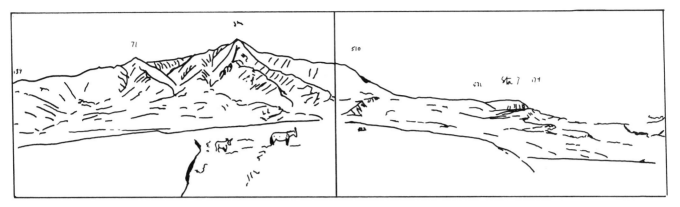

Figure 10.8. View west to Mt. Pennell. The ridge marked "Sta. 7" is the Horn laccolith; "511" is the Dark Canyon laccolith. The Mt. Pennell stock forms the peak and extends southward to Straight Creek, in the canyon emerging from the mountain on the north side of Sta. 71. Between stations 71 and 164 is Bulldog Ridge.

Top of lower sandstone at 11.40 = 23.41 = 7650.

Sta. 44 = Dinner Camp = bluff of Lower sand = 1.30 = 23.35 = 7720.

The dip of the lower sandstone measured is gently toward Pennell and all these salients (sta. 44 etc) but a little further up the slope it changes and becomes about the same as the slope. Nothing appears above the lower sand on the road to the mountain and the exposures near 512 are the same bed rising toward the Pennell-Maze fold.

{p. 15}   Camp 26 is the Pennell camp of last year. At 7.10 21.02 = 10,580. Sunset 5.35 pm The Henry Mts cast their shadow against the La Sal.

Friday, November 3. Camp 26 at 9.25 = 21.03 = 10570.

Camp 27 (in the aspens at foot of Pennell) 11 am = 21.03 = 10570. The camping ground has been occupied before and we read on the trees

T. Printz   1875

L.C.J.   1875

T. J. Howard Sep. 1 T H E 30.   75

C. Bau er

Christ our Father

        L O R

        T

        A II

            4

            (1

A.S. Parmelee

            1875

                Sep. 31

S S

A F P

        Sept. 30,   /75

{p. 16}

R I C K P

E I

T.J.H.   1851 (a fraud)

S S Graham   1875

{p. 17}   [*missing. Continue with back of p. 17.*]
variable (chiefly uphill) during yesterday p.m. and this a.m., but last evening it was all down hill.

At camp 26 the wind was W (down the mt) as long as we stayed (5 pm to 9 am).

The three inches of snow in Camps 26 and 27 we found represented by 12 on the higher parts of Pennell. We made the climb of 2050 ft in 2 hrs 25 min.

Camp 27 1.5 pm = 21.04 = 10,550

Camp 9 at 1.23 pm = 20.80 = 10,880

Camp 19? on Pine Alcove Cr 3.35 = 22.90 = 8250

{p. 18}   In descending to camp 19 I find that much of the upper part of the slope is floored by a sheet and that the small lacune west of Sentinel Butte is an affair of some magnitude the debris from which is of large round massive boulders like those from the Sentinel.

Divide 4.10 = 22.69 = 8520

Foot of Gate Cañon V.S. [*Very South*] Cache. Top of Gate sandstone = 1.45 = 22.97 = 8170

Sunday, November 5. Camp 28, 7.35 pm (?) 23.27 = 7800

The lowest point on our trail down V.S.C. Wash reads 23.66 = 7430. When we reach the top of the Gate sandstone on South Cache Cr. I read 9HO = 23.20 = 7890. There is a coal seam here with 30 ft of sandstone cover. Its floor is shale—2 ft thick and then more sandstone. Coal 4 ft. Dip 15°.

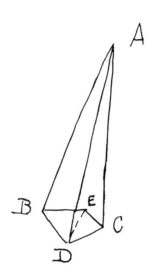

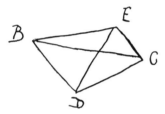

Figure 10.9. A = Sta. 20; B = 214A; C = 213a; D = 213c
Known angles are BAC, BDA, ADC, DCB, BDC. Required BAD.
The angles at A are so small that they may be considered proportional to the sines.

$$\frac{BAD}{CAD} = \frac{SinBAD}{SinCAD} = \frac{SinABC:SinBDA:SinDCB}{SinACB:SinCDA:SinDBC}$$

I mistake—Before this can be done the angles at B and C must be derived from those at D and E the stations occupied.

ED:DC::ECD:DEC
BD:ED::BED:EBD
BC:BD::BDC:BCD   sines all
DC:BC::DBC:BDC

$$\therefore \frac{SinDBC}{SinBCD} = \frac{SinDEC}{SinECD} = \frac{SinEBD1}{SinECD}$$

The second equation is DBC ≠ BCD = 180° - BDE
is where I do not know how to continue.
The simplest solution is by plotting.
[*These calculations on p. 16 evidently are an attempt to solve some un-explained problem in Gilbert's triangulation. Most of the entry seems to have been crossed out with two very light lines.*]

{p. 19}   Camp 1 at 10.50 = 22.78 = 8300
At 1.45 pm = 22.80. Marched to camp 17 at the foot of the Hilloid Butte.
Monday, November 6.
Ice in the kettles at Camp 17. 7.30 = 22.48 = 8750
Climbing Ellen we ride to 8.35 = 20.87 = 10,770. Then we skirt about the flank of pt. 5 to 9.15 = 20.80. The valley we have followed is choked with snow-slidden debris. Make out no struc-ture to pt. 5 [*spur north of Mt. Ellen*] but the slope is of shale and trap. Some of the shale is little altered but more of it is slate.
Rising however to the 5-1 saddle [*north side Ellen Peak*] 20.50 = 11,260 I find all shale (below) and slate (above) except a narrow dike along part of the divide. At 9.55 = 20.25 = 11,600 I project tree tops on pt. 8 against the horizon but am not yet up to the level of 5. There is the same shale zone judging by vegetation etc on the 8-1 saddle [*northeast of Ellen Peak*]. The persistence of the zone so far about the flank of Ellen is a very notable feature and as yet unexplicable.
5 is even with the horizon at 20.08 = 11,840. 26 and 2c [*a mile NW of Ellen Peak*] are higher.

{p. 20}   We rest at the top of the timber (on this line) 10.35 = 19.83 = 12,180.
Sta. 1 = Ellen Peak, 11.10 = 19.38 = 12,130
                    1.22 = 19.33
Descending I reach the level of 2c (30 ft higher than 2b) at 19.86 = 12,130. 2d at 20.30 = 11,550.
The Hilloid shows by colors that sedimentaries lie against it on the side toward pt. 5. The profile seen from the south confirms that seen from Sta. 2. There is a fine easy arch to the NE to complete, with the steep one on the West, a simulation of Mt. Hillers. Camp 17 again 4.25 = 22.50 = 8730.
Tuesday, November 7. Camp 17 at 8.27 = 22.50 = 8740
Homeward bound. Election day.
at 9.40 = 23.45 = 7620. 'We have' many outcrops of lower sand near end of N Butte.
See p. 22 for sketch of the Hilloid Butte. It is buttressed all the way around and the lower sand can be (p. 21, Fig. 10.10; p. 22, Fig. 10.11)

{p. 23}   seen rising against it from the east. Not less than 2 m *N* of the butte the same sand shows with a dip of 8° or 10°. It is about on my level, 24.17 = 6780. We reach the Dirty Devil River at 3.40 at the base of the Gate Shale and at the top of the coal of the lower sand.
There is a very curious phenomena of badlands across the river. At a score of places I can see them vertical on the NE side and sloping on the opposite. The prevailing wind must be from the side that has the slope. Can that in some way account for the phenomenon?
This Dirty Devil drift contrasts strongly with that of the Henry Mt. benches. All the trachyte is well rounded and all of it is vesicular. 'There' is much chert, Shinarump wood, jasper etc. and some limestone.

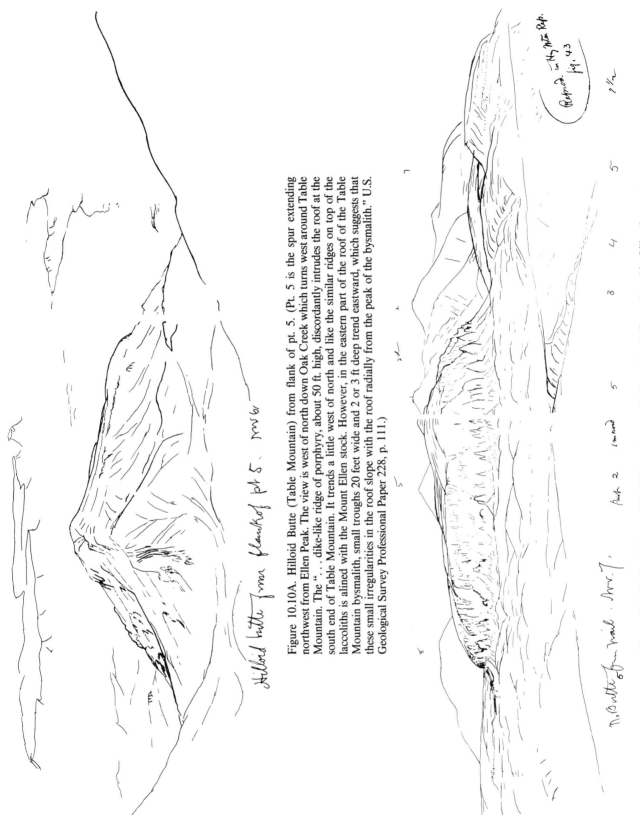

Figure 10.10A. Hilloid Butte (Table Mountain) from flank of pt. 5. (Pt. 5 is the spur extending northwest from Ellen Peak. The view is west of north down Oak Creek which turns west around Table Mountain. The "... dike-like ridge of porphyry, about 50 ft. high, discordantly intrudes the roof at the south end of Table Mountain. It trends a little west of north and like the similar ridges on top of the laccoliths is alined with the Mount Ellen stock. However, in the eastern part of the roof of the Table Mountain bysmalith, small troughs 20 feet wide and 2 or 3 ft deep trend eastward, which suggests that these small irregularities in the roof slope with the roof radially from the peak of the bysmalith." U.S. Geological Survey Professional Paper 228, p. 111.)

Figure 10.10B. North Butte = Hilloid Butte, from trail, November 7. Reproduced in Gilbert's report as Figure 43.

Figure 10.11. Sketch of the NE Butte (Bull Mountain) as seen from bench near the Dirty Devil River. November 7, p. 24 of Notebook 10.

Camp 29. 4 pm = 25.01 = 5830. Fuel furnished by a standing cottonwood killed by beavers.

Wednesday, November 8. Camp 29 at 9.15 am = 25.14 = 5700

We follow the badlands between Needle Butte [*Factory Butte*] and N Twin Mesa [*North Caineville Mesa*] and reach the lower sand at

{p. 25}   24.66 = 6230. The top is 200 ft higher. The dip is 33°. The strike of the Gate fold here curves convex to the east.

As I sit on the apex of the lower sand I am on a level with a line 200 ft above the Red Vag. which is rain-buttressed with great beauty.

The Rock Hills Swell has a node of increased flexure where the Gate fold joins it. East of this node the max dip is only 15° and SW about the same. In the node it may reach 45°.

The great *"Inoceramus"* is the mark of the Gate shale in this region. In crossing badlands today we have seen thousands and the max size is over 3 feet and perhaps 4 ft.

Yesterday we crossed the Indian trail twice. They returned westward close to Hilloid Butte with 26 animals and at least 6 prs of moccasins. They passed eastward in two parties (one earlier than the other) crossing Cache Creek near the South Twin.

Camp 30 (12 m = 24.72 = 6180) is on the lower blue shale on the line of the Blue Gate fold. A little water rises (temporarily) in a wash which cuts the Lower sand hogback and makes for Curtis Creek. Not a prominent spring.

{p. 26}   *Thursday, November 9.* Camp 30. 8.45 = 24.73 = 6150

We quickly come upon the Belted Cliff. The pale band (A in Fig. 10.12) has a firm sand at *a* which we find passable and follow from 9.5 = 24.65 = 6250 to 10.25 = 24.90 = 5950 where we reach a large salt stream which I take to be a tributary to Curtis Creek. It carries M.L. trachyte in small quantity, is very salty and has no trees nor canes.

Figure 10.12. Cross section of the impassable escarpment of the Belted Cliff crossed at the Salt Creek November 10.

Sand ripples by wind 15 inches apart.

We stop for lunch on the salt stream where the trail next touches it. It is nearly fresh.

The red and green shale is saturated with gypsum which lies in reticulated veins (Fig. 10.13), which for the most part cross the bedding at small angles. The gypsum is fibrous and the fibers are approximately vertical whatever the angle of the vein or the lamination.

{p. 27}   Noon camp 2.30 pm = 24.71 = 6170

The whole Belted Cliff below the green band is rain-fluted in most beautiful style and garnished with buttresses—flying buttresses and spires—beyond the power of my pencil to express. These are fine subjects for the camera. The red of the cliff is chocolate and identical with the red of the gypsum beds, but the green of the cliff is grayer than that of the gyps-shales.

In the distance I see some of the problematic buttes of the synclinal with sheets of lava.

After dinner I soon come to a dike (A) standing vertical and seen through the whole height of the cliff—not less than 2000 [*200?*] ft. It thickens from 10 to 20 ft and the material is

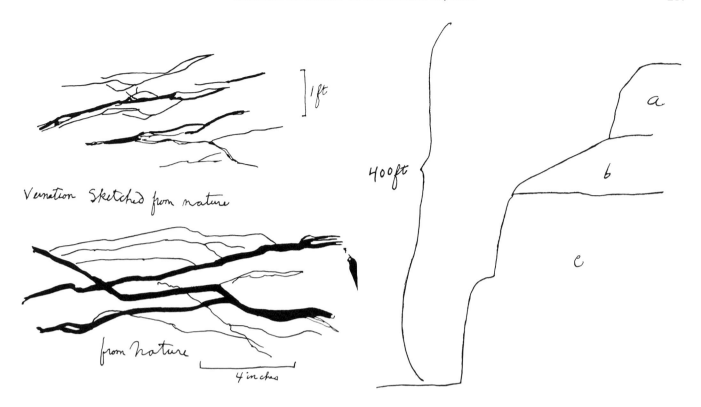

Figure 10.13. Reticulated gypsum veins in the Belted (November 9).

Figure 10.14. Section of the Gray Cliff Sandstone and overlying beds in cliff examined November 10.
"a = Calcareous Jura; gyps at base
b = reds and soft
c = Gray Cliff Sandstone"

{p. 28}   basalt. No sheet nor 'coulee' is visible at top.

At Camp 31 = 5 p.m. = 24.60 = 6300. There are a number of other dikes, thin, vertical, as straight as joints usually are, basaltic, traceable for long distances, trending about N/S and in one instance reaching the gypsum beyond the wash.

(B) The dike at camp is six feet thick and trends S 55° (mag). The red vag adjacent to it has its color discharged for 3 ft. and is somewhat hardened beyond. Beyond camp is what I take to be a coulee of basalt on top of the Belted Cliff. Its undulation and the absence of other rock above it mark it as a *coulee* rather than a sheet.

Friday, November 10. Camp 31, 8.10 am = 24.45 = 6470

We take trail up the top of the Jura. At 8.40 = 24.03 = 6930 I am not half way to the level of the top of the cliff capped by the basalt flow which is above the

{p. 29}   upper red of the Belted. The section here is (Fig. 10.14).

At 23.60 = 7430 I am near the summit of the trail (across the salient of the Rock Hills Swell) and am nearly as high as the coulee near camp. No I think that the lavas I took for *sheets*

yesterday are coulees. What I supposed a cap of sandstone is really a cap of lava. The erosion had progressed to various deposits in the vag, when the lava overflowed. In the same region the drainage is now in the foss. Jura and below. [*Gilbert refers to these igneous masses as "lava." They are part of the field of analcite diabase sills and dikes mapped and described in Gilluly, 1927 and 1929 (see References in Introduction). In the next day or so Gilbert conclusively shows they are in fact intrusive; Figs. 10.16 and 10.17.*]

At 10.15 = 23.36 = 7700

{p. 30}   We are still on the gently sloping foss Jura and are higher than the coulee near camp 31.

Surely I see no sheets that I can be sure of; but many dikes.

We find a little snow on the S sides of hills and water our horses.

In the Jura carb sands are ripple marks, 13 inches from crest to crest and 1⅝ inches high.

The other day I measured an *"Inoceramus"* 43 inches across.

Later we approach buttes that surely are filled with sheets of intrusive rock, even and broad—at 3 or 4 horizons and flooring terraces. The kind of rock remains to be ascertained.

Two dikes that we touch are of the same basalt gathered yesterday.

The general trend of dikes is NW (map).

Camp 32 at 4.40 pm = 23.40 = 7650. We find difficult water in a cañon of the Jura foss. *"Campetonectes"* shows feebly.

{p. 31} A dike of basalt in the cañon 2½–4 ft thick I see not visibly affect the limestone at contact.

The angularity of the sheets in the buttes we are to pass tomorrow is shown on the next page (Fig. 10.16).

Saturday, November 11. Camp 32 = 8 am = 23.20 = 7900

Of the specimens gathered at the pass today, those marked (A) are undoubtedly from sheets and those marked (B) are probably so but possibly from dikes or a coulee. I think the top lava sketched on p. 32 (Fig. 10.15) is a coulee.

Where I doubted yesterday whether the upper bed of some butte was lava, I am now convinced that it is not for I find that each terrace due to a sheet carries a cap of *altered* sandstone on top of the basalt. But the lava bed near camp 31 and the one on top of the diagram above, have no such caps. Hence, I call them coulees and say that the deepest buried sheet I have yet seen was intruded below only 1000 ft of sandstone.

Trend of dikes NW and NNW.

Yesterday I saw a little of the 'crapping' [*contorted bedding?*] Powell has told of in the walls of foss Jura.

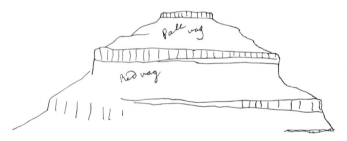

Figure 10.15. Diagram of the sheets in the butte near camp 32.

{p. 33} It extended in only a few feet.

As a rule the rock containing the sheets is altered for a few feet only (judging by color and habit) and most where the sheet is thickest. I think the Sheet Island at the pass shows a depth of 75 ft. Think none exceeds 100 ft.

I can trace one dike through two sheets, from one of which it is distinguished by color. Other dikes I see connected with sheets. Figure 10.16 with sheets 15 ft thick is plain.

The sheets all show prismatic cleavage running from both sides to the middle. The sheets are marked as subsequent and not as contemporaneous by the metamorphism of the superior bed.

Pass at 10 = 23.68 = 7920

(c) Descending we find a thin sheet on top of the gypsum.

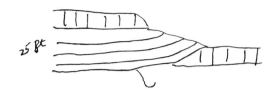

Figure 10.16. A well-exposed section with sheets 15 feet thick, referred to on p. 33.

{p. 34} A fine proof of the subsequent date of the sheets is shown by the phenomenon sketched in Figure 10.17. Where the red Vag. cleaves along a broken line for the admission of the lava, discoloration has advanced much further *along* the lamination (at the broken ends) than *across* it.

At dinner camp we are not yet out of the land of dikes and one at least bears ESE (m). Nor are the sheets confined to islands of Vag. I see a broad thin one on a peninsula.

Dinner camp at 12 = 23.02 = 7420 is rather low in the red Vag. There is a gentle dip all through the region from the Rock Hills. I measure dips ranging from 2° to 8°.

The water we dine on is the best since the Dirty D.R. but is not quite sweet. The boys are all physicked by the bitter and briney water we have had to use.

We fail to reach Carter's Creek as intended and encamp again in the Vag on a dry wash that probably leads to the creek near the head of its cañon in G.M.

Figure 10.17. That the sheets are intrusive is shown by the discoloration extending along laminae.

{p. 35} The table E of our dinner camp has 1 or 2 sheets of lava near the middle of the red Vag. and east of it are sheets above or in the gypsum, while dikes stretch nearly to camp 33.

The sheet which I sketched at top of p. 34 where it changes its horizon has the same double jog more than once and besides has a general descent in the strata toward the NW as though it had tried to keep at one depth below the existing surface when it was intruded.

Sunday, November 12. Camp 33 at 8.45 = 23.82 = 7180

In our march we travel chiefly on a heavy sandstone due to a hardening of the gray-green band of the Belted Series. I think the conglomerate above is very thin.

Elisha kills a deer at dinner camp. (Yes, two.) 12.45 = 23.80 = 7200.

{p. 36}   Camp 34 is at the top of the upper Red Vag and reads 23.80 = 7200 at the top of the conglomerate. Close by I read 23.50 = 7540.

Monday, November 13. Camp 34 at 9 am = 24.05 = 6910

At 11.30 we are on the top of the *"Gryphaea"* bed, identifying by the white and purple shale rather than by fossils. There seems to be no lower blue cliff.

Camp 34 is at the foot of the cañon through sandstone *B* as I will call it for the present and is only 100 feet lower than the lowest sandstone. In the sandstone near the base are oysters, smaller than the *"edulis"* (coll).

Pangwich rolled over today into Curtis Creek. This is her third roll on the trip. Beck has accomplished two and Gomas, Joel, and Lousey one each. Our little train of 9 animals has attained to seven rolling scrapes.

{p. 37}   **Tuesday, November 14.**

For two nights I have had an eye out for a shower of meteors but without success. [*Gilbert evidently was familiar with and looking for the Leonids.*]
Camp 35 at 7.50 = 23.73 = 7280

A special search fails to detect in the *b* shale any of the great *"Incoceramus"* which mark the Gate Shale and are absent from the lower in the Henry Mt. region.

I think I have never noted that a synclinal which we crossed on the p.m. of the 11th joins the R.H. Swell [*Rock Hills*] nearby and makes a reentrant in its line of maximum flexure.

Climbing the cliff is section in Figure 10.18.

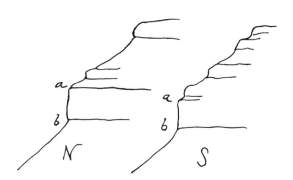

Figure 10.18. "Climbing the cliff— b) Base of sandstone 23.54 = 7500; a) Top of first bed of sand 23.45 = 7608.
The distance measured being *a-b* in each profile, these are the alternations of sand and shale which I see on the N and S sides of the Ivie Creek cañon."

{p. 38}   The mass of Castle Valley Cliff which I noted last year is surely a slide. It is itself divided and dips to the NE (the parent cliff to the W). This might be if it was faulted but it lacks the buttress of the next lower order which it would have in that case. It has slidden over and buried the lower cliff. I know no slide so massive. It is more than a mile long and ¼ mile broad.

So much for first impressions. Two hours later I have crossed an old terrace of Ivie Creek and command the valley. Now I can see that my "slide" is surely a fallen block in a zone of inverse displacement. The zone of faulting runs from the "slide" to the outlying butte N of Ivie Creek and is marked all the way by an inner cross-valley (*a*) (Fig. 10.19).

{p. 39}   Later views confirm the last, but there is nothing clear except there have been faults. Ivie Creek exhibits some beautiful veneered benches—not less than four terraces above the flood plain.

The coal I see nothing of. That and the faults would make an interesting study together.

Camp 36 is on Ivie Creek below or at the lower margin of the zone of faults. All the rock is homogeneous gray shale but there is sandstone ahead—not far.

Wednesday, November 15.

Camp 36, at 9.45 = 22.96 = 8180. Wagon road and houses at 11 am = 22.75 = 8440. Gilson's Ranch at fork of Ivie Cr 12.35 = 22.37 = 8900 (20 ft above water). Pass 21.82 = 9560. Camp 37 is on the head of the water in Meadow Cañon, Howell's Hole.

{p. 40}   **Thursday, November 16.**

Mr. Sam Gilson yesterday found me a mare (colt following) to use to Gunnison, the agreement being that his men are to bring Gomas there within three weeks and I am to leave the mare and colt there in good hands for him to take again.

Today we march from camp 37 near the divide to camp 38 below the creek in Salina Cañon.

Friday, November 17. Camp 37 to Salina.

Saturday, November 18. Salina to Gunnison

Sunday, November 19 and

Monday, November 20.   Gunnison

Tuesday, November 21. Gunnison to Nephi. There can be no doubt about the fault along the base of the San Pete Plateau.

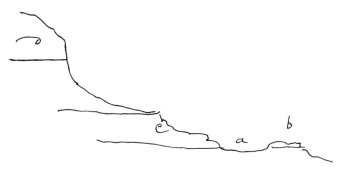

Figure 10.19. "I don't identify *b* yet with either *c* or *d* but it might even be B (p. 36). The dip in *b* is diverse."

{p. 41}   **Wednesday, November 22.** Nephi to Salt Lake City.

[*This and the following pages (p. 42–45) pertain to Lake Bonneville and are given in my account of those notebooks (see Gilbert, 1982 in References in Introduction).*]

{p. 46}   *Approx thicknesses*
    Upper cliff   1000 ft
    Gate Cliff   1500
    lower Cliff   700
    conglom and vag   1500
    Gray Cliff   500
    Verm cliff   400
    Shin. shale   600

{p. 47}   [*Fig. 10.20, an unlabeled sketch on p. 47; some alluvial terraces.*]

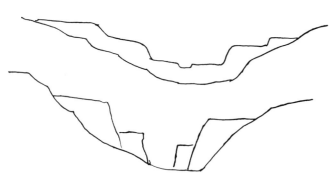

Figure 10.20. Unlabeled sketch on p. 47 showing some alluvial terraces, probably inspired by Ivie Creek; see Notebook 10, p. 39. The lower diagram probably represents some speculation about the stratigraphy of the terraces.

{p. 48}   S. H. Gilson, Nephi, Juab Co., Utah aeronautics
Mark Tully, White House, Salt Lake City
F(ranklin) L. Farnsworth, Kanab, Ut
Kane A.N.
Elisha Averitt, Kanab, Kane Co., Ut
Nathan Adams, Kanab

{p. 57}

Gunnison	Salt Lake
Telegram	Bank
Mail	Letters
Gilson's horse	Reading matter 'Mga'
Box specimens	Plotting
	Bath, hair, drawers, collars, gloves
	Clayton

{p. 55–57}   [*Lists outfit for camping; see p. 20 of my account of his Lake Bonneville notes.*]

{p. ?}   Salina Nov. 17 Bot. of Mr. Martin
    Hay for 9 horses at 25   2.25
    6   wheat 1½   1.00
    2 lbs butter   (25)   .50
    1 doz eggs   (25)   .25

{p. ?}   Thompson's account with GKG

Dr. ' '	22.98	Cr. on vouchers (field ex)	19.00
From Ogden to S.L.	2.00	cash to men	10.00
cash	300.00	on Salina vouchers	4.00
	324.98	Cash	291.98
			324.98

APPENDIX A. GILBERT'S CAMPS

1875

1. At York, Utah, June 20
2. Nephi, June 21
3. Taylor's Ranch, about 20 miles south of Nephi, June 22
4. Gunnison, June 23–25
   5 and 1 again. At Salina through June 30
2. Salina Canyon, July 1
3. Salina Canyon, July 2
4. Salina Canyon, July 3, 4
5. Meadow Canyon, July 5, 6, 7
6. Fish Lake, July 8, 9
7. Rabbit Valley, July 10
8. Supply Camp, at the Red Gate, July 11–13
9. Southeast of Teasdale?, July 14
10. SE side, Thousand Lake Mountain, Corral Creek ?, July 15, 16
11. White Crag (= Cockscomb) between Teasdale and Grover, July 17
12. Edge of G.M. near Square Butte (= Lion Mtn), July 18
13. 2,000 ft below rim at east salient of Aquarius Plateau, July 19
14. Foot of east salient of Aquarius Plateau, July 20
15. In tributary of Pleasant Creek, at first bench below rim of the Aquarius Plateau, July 21
16. Bench below SE salient of Aquarius, July 22
17. Boulder Creek, above Navajo Sandstone, July 23
18. Near top of Aquarius at head of Boulder Creek, July 24
19. In G.M. at top of Aquarius, head of Pine Creek, July 25
20. Bottom of valley of Pine Creek, July 26
21. East of camp 20, south side Escalante River in vicinity of False Cr., July 27
22 to 28. Along Straight Cliffs of Kaiparowits Plateau, July 28–August 6
29. On Aquarius near divide to Rabbit Valley, August 7
30. Same, August 8
   Return to Supply Camp, August 9; remain there 10–11
31. Along faulted south edge of Teasdale anticline, August 12
32. Continue along faulted SW side of Teasdale anticline, August 13
33. On waterpockets of G.M. a little south of Pleasant Creek, August 14–15
34. Sorenson Pockets, August 16
35. Hall Creek 7½ hours march south of camp 34, August 17
36. Waterpocket Canyon, August 18–20
37. Adams's waterpocket, August 21
38. SE base Mt. Hillers, August 22
39. Below Shinarump at W base of Ellsworth, August 23
40. Trachyte Creek or Straight Creek, August 24, 25; West Camp, by Straight Creek, August 26
41. East base of Mt. Pennell, north of Straight Creek, August 27
42. Along Summit of Mt. Ellen; at Bull Creek Pass ?, August 28
43. "On villainous creek tributary to Dirty Devil," August 29
44. Along Dirty Devil east of Twin Mesas, August 30
45. Along Dirty Devil just below the canyon through Capitol Reef, August 31
46. Along Dirty Devil at head of the canyon, September 1; Return to Supply Camp, September 2

47. On shoulder of M.L., September 3
48. NE corner of M.L., September 4 (Thousand Lake)
49. Near the salt lake east of Mt. Hilgard, September 5
50. Forks of Salina Creek September 6, 7
51. Limestone ledge below Musinia Peak, September 8
52. Twelve Mile Creek, September 9
   Return to Gunnison, September 10

1876

1. Cache Camp, by Dugout Creek, at foot of Mt. Ellen, September 9, 10; returned there Oct. 14–16
2. By Dugout Creek at mouth of Pisto Creek, September 11
3. Summit camp, at spring on southeast side of Ellen Peak, September 12, 13
4. By Bull Creek at foot of Sawmill Basin, September 14–16
5. By Crescent Creek on the gravel benches, "Pleasant Camp," September 17, 18
6. Avarett Spring, in Copper Creek basin, September 19, 20
7. Garden Basin Creek, south of Copper Ridge, or Slate Creek, September 21
8. Summit Spring in Pennellen Pass, September 22, 23
9. North slope Mt. Pennell near west rim of Dark Canyon, September 24
10. In Cretaceous shale by Straight Creek, September 25
11. South Pass, September 26–28
12. Fork of Trachyte Creek heading at Hillers Peak, September 29, 30
13. Trachyte Creek just above canyon in Navajo Sandstone, September 30, October 1
14. Near Starr Spring, October 2
15. Pine Alcove = Bullfrog Creek at Ferron Sandstone, October 3, 4
16. Divide between Bullfrog and South Creek above Steven Narrows, October 5–8
17. Base of Hilloid Butte = Table Mountain, October 9–11
18. Base of Pt. 313, near Table Mountain, October 12, 13; Cache Camp, October 14–16
19. Half mile upstream from camp 10, October 17
20. On Dinah Creek below Dinah Ranch, but editor not sure what is at Dinah Ranch—a grassy meadow? October 18
21. Base of Pt. 223 on southwest side of Mt. Pennell, October 19, 20
22. Upper Dinah Creek at base of Mt. Hillers. October 21
23. Base of Mt. Ellsworth (Fig. 9.10), October 22–27
24. West base Mt. Holmes, October 28–30. And to camp 14.
25. Trachyte Creek, October 31 and November 1
26. At Pennell camp of 1875, November 2
27. Camp 9, November 3
28. North side of Ellen Peak?, November 4
29? Oak Creek springs area, November 5
   Base of Table Mountain, November 6
   Heading for home, November 7 (see Fig. 10.11)

221

APPENDIX B. GILBERT'S INSTRUMENT STATIONS

Gilbert's 1875 instrument stations were mostly west of the Henry Mountains.

1876

1. Ellen Peak, correct altitude 11,506
2. Bull Mountain = NE Point, correct alt. 9,187
3. Wickiup Ridge, correct altitude 9,760
4. Copper Ridge, altitude 8,950
5. The Block, alt. 7,646
6. S end Mt. Ellen summit ridge, spur at 9,590 W of head of Slate Cr.
7. The Horn = Pass Butte, alt. 9,047
8. Ragged Mountain, alt. 9,113
9. Mt. Pennell peak, alt. 11,371
10. Probably Brown's Knoll of topographic quadrangle map, alt. 8,149
11. Probably Bulldog Peak of topographic quadrangle map, alt. 8,165
12. Easternmost of the high peaks on Mt. Hillers, alt. 11,700
13. Mt. Hillers peak
14. On Taylor Ridges ?; pt. 87, Gilbert aneroid 6,710; alt. 5,500
15. Near Trachyte Creek ?; Gilbert aneroid 7,310; alt. 6,000
16. Trachyte Creek N of Trachyte Mesa = pt. 250
17. SE of Black Mesa, alt. 6,520
18. No station 18 found in the notes
19. Fred's Ridge = pt. 217; this station number used again on Ellsworth
20. Swap Mesa = pt. 159

21. = pt. 74 on ridge of Ferron Sandstone (possibly a sill), on NW flank Pennell
22. = pt. 105 on east side of Tarantula Mesa
23. Between pt. 105 and 104
24. Table Mountain
25. Knob ½ mile NW of Willow Spring
26. Knob on ridge of Morrison Formation on Cedar Creek benches
27. = pt. 240; knob on Dugout Creek fan
28. = pt. 27 = old gravel covered pediment along South Creek
29. = pt. 26a; NW base of Mt. Pennell
30. pt. 223
31. North Summit ridge of Mt. Ellsworth
32. = pt. 385, W of Sta. 31; on Shinarump; Fig. 9.18
33. SW of 32
34. On Ellsworth, in trap along trail E from Camp 23
35. ?
36. One of summit ridges on Ellsworth
37. ?
38. ?
39. = pt. 213 on W ridge of Mt. Holmes
40. Head of spur 498
41. ?
42. Spur ? on N side Mt. Holmes; pt. 258
42. (Number duplicated); = pt. 258
44. Trochus Butte

APPENDIX C. PRINCIPAL SURVEY POINTS SIGHTED AND USED BY GILBERT IN HIS DESCRIPTIONS OR DRAWINGS

1875

Most of Gilbert's 1875 points are west of the Henry Mountains. Point 119, which is referred to, is the conspicuous stack of Wingate Sandstone at the southeast end of the Teasdale anticline.

1876

2. Peak at Deer Heaven, a mile NW of Ellen Pk.
5. Dry Lakes Peak, a mile N of Ellen Peak
6a, 6b. In Cottonwood Canyon draining NW from Ellen Peak
7. Table Mountain = Gilbert's Hilloid Butte
8. Peak of Horseshoe Ridge
9. East end of Horseshoe Ridge
10. Birch Creek Canyon above Bacon slide
11. Bull Creek, at and below Sawmill Basin
12a, b, c = Wickiup Ridge
14. = Station 2; Bull Mountain; Gilbert's NE Butte
15. = Station 3; highest part of Wickiup Ridge = 12a
16. Point of spur north side of mouth of canyon draining Bromide Basin
17. Barton Peak
18. Granite Creek
19 and 20. North Summit Ridge of Mt. Ellen
21. Bull Creek Pass
22. South Summit Ridge of Mt. Ellen
24. Granite Ridge
26. South Creek Ridge?
27. Old gravel-covered pediment along South Creek
29b–e Along Dugout Creek and its north fork

34 to 38. Points along the escarpment forming the east side of the Crescent and Maze Arches
39. Station 5, The Block in the Maze Arch
40. = Station 8; Ragged Mountain
41–44 Omitted
45–49 Poison Spring benches and others east of Granite Creek
50–54 Escarpment of Morrison Formation extending from Poison Spring benches to Trachyte Creek
56–58 Copper Ridge and the point north of it
59. Black Table = Black Mesa
60–62. Gilbert's Jerry = Cockscomb or Sawtooth Ridge
63–69 Peaks on Mt. Hillers
71. Bullfrog Ridge, overlooking Straight Creek
72 and 73. Pennell Peak
74. See Sta. 21, Pine Spring
75. Sta. 7; The Horn
76–79 Spurs on south end of Mt. Ellen, but the numbers appear duplicated along the escarpment of Emery Sandstone 4 miles west of Pennell Peak

From here on, some numbers are omitted.
87. Sta. 14
91–94. Circle Cliffs
95–98, 101. Emery Sandstone along Waterpocket Fold
99 and 100. Waterpocket Fold
102–106. South rim, Tarantula Mesa; 105 = Sta. 22
112–126. Emery Sandstone escarpment at Stevens Narrows and north to Steele Butte; 123 = Steele Butte; 114 = Camp 16, divide between South and Bullfrog creeks

127–129. Old gravel-covered pediment along South Creek

135. No Man Mesa

136–144. Rim of Emery Sandstone along Bullfrog Creek below Scratch Canyon; junction Bullfrog and Scratch Canyon

145–147 and 150–154. Cave Point area west of Bullfrog Creek

149. Trachyte Mesa (see p. 8.1)

150, 151, 155–158, 166. Rim of Emery Sandstone along west side of Muley Creek and west to The Post

159. Gilbert's "Point Retroussie" = Emery Sandstone tip of Swap Mesa

164. Bulldog Ridge

172. Stewart Ridge laccolith

177, 178, 180 on west base of Hillers

183. Peak on Hillers

184, 187, 189. Peaks on Sawtooth Ridge

190, 192, 193. On pediment on east side of Mt. Hillers

197–200. Rim Navajo sandstone along Trachyte Creek, Fig. 7.31

201–205. Run along N side of 2-mile Canyon, Fig. 7.31

213–215. Mt. Holmes, summit dikes

216. Gray Cliff on Mt. Ellsworth

217. Freds Ridge = sta. 19 = 403

219. Mt. Holmes

220. Mt. Ellsworth

222. Mesa above Hite

223. 9345 on SW flank Mt. Pennell; number duplicated on Ellsworth

229. Mesa between 4 Mile and Ticaboo Canyon

240. Bulldog Peak (of Professional Paper 228)

241–245. Points on south rim of Lower Leaden = Dakota Sandstone

247. east of Jerry, a knob along Sawtooth Ridge

248. sill east of Black Mesa

249. Trachyte Mesa; see Gilbert's Notebook 8, page 4.

250. Gravel-capped butte by Trachyte Creek

254. Pulpit Arch area

255. Spur(?) on N. side of Mt. Holmes

256. Turkey Knob = Trochus Butte

257 and 260. Crags and buttresses at SE base of Mt. Hillers

269½. Axis of syncline between Mts. Holmes and Hillers

271a. N side Mt. Holmes

273. Dike on Mt. Holmes

275. Navajo Sandstone ridge on west side Ellsworth; see Fig. 8.6

276–280. Crags along the south base of Mt. Hillers

281–283. Crags at east base of Mt. Hillers

285. Northwest spur of Mt. Ellsworth (?)

304. Escarpment above 305

305 and 306. Escarpment of Emery Sandstone along SW flank of Pennell

311 and 312. Ridges on the north and south sides of Cedar Creek

313. = Sta. 26

314–316. Mostly Ferron Sandstone at W base of Mt. Ellen but may include Morrison Formation in dome at Cedar Creek

317. Cedar Creek laccolith

318–323. Northeast side of Stephens Mesa

327. Near north end of the Cedar Creek laccolith

330. Apex of Cedar Creek fan

334–336. On Pistol Ridge

337 and 338. Cedar Creek laccolith and gravel bench

340. North Cedar Ridge

341. Gravel-covered pediment on Cedar Creek laccolith

342–348. South flank of the Cedar Creek laccolith

350 and 351. Dugout Creek at and just above the canyon mouth

356. (Probably 386). West spur on Mt. Ellsworth, at edge of stock?

357 and 358. Dugout Creek laccolith, Sarvis Ridge

359. West end of Sarvis Ridge

373. East side of Mt. Hillers

379. Dakota Sandstone west of Mt. Holmes

384. unidentified point on Mt. Ellsworth

385. = sta. 32

387. West spur Mt. Ellsworth

388. West spur Mt. Ellsworth

392 and 393. North Summit Ridge of Ellsworth

396. North spur of Mt. Ellsworth

399. One of Ellsworth summit ridges

402. East summit ridge of Mt. Holmes

403. = 217

416. 2 Mile Canyon

423. S rim 4 Mile Canyon

497 and 498. Buttes on N side Mt. Holmes; 498 = Sta. 40?

500. N slope Mt. Holmes

504. North slope, Mt. Holmes

505. E edge of gravel bench S of Pulpit Arch

510 and 511. Dark Canyon laccolith

Page numbers prefixed by "0" refer to Introduction. Those prefixed "1" to "10" refer to Gilbert's notebooks, and the pages are those in the notebooks and not pages in this book. Numbered figures are by Gilbert; figures identified by letter are additions by the editor.

Typeset by WESType Publishing Services, Inc., Boulder, Colorado
Printed in U.S.A. by Malloy Lithographing, Inc., Ann Arbor, Michigan